AMERICAN CYBERSCAPE

AMERICAN CYBERSCAPE:

Trials and the Path to Trust

MARI K. EDER

Defense
Press

ISBN: 978-1-953327-00-0

Published by Defense Press
www.defensepress.com

CONTENTS

FOREWORD

Trust is the glue of life. It's the most essential ingredient in effective communication. It's the foundational principle that holds all relationships.

—Stephen Covey

Trust is very hard if you don't know who you're trusting.

—Marianne Williamson

These quotes summarize the dilemma facing most of us today. We want to trust others. We want to trust those providing information to us. But, with the avalanche of information produced on a daily basis often from sources with a distinct bias in their reporting or from sources that are hidden from us through layers of innocuous sounding organizations or outright subterfuge, it has become incredibly difficult to know who to trust and what is accurate. Once information is "out there" via today's instantaneous technology, it is nearly impossible to contain it or correct inaccuracies.

Make no mistake about it, there has always been some lack of trust in what the media and elected officials say; it goes with the territory. But there was a time, and not all that long ago, when the media and elected officials were held in much higher regard than they are today. Reputation and trust go hand in hand. A 2018 Knight/Gallup survey found sixty-nine percent of U.S. adults have personally lost trust in the news media in the past ten years. A July 2019 survey by the Pew Research Center found seventy-five percent of adults have little or no confidence in the federal government and sixty-four percent believe we have little trust in each other. It is not just the institutions of media and government that have a reason to be concerned by these results; communicators at every level and for every organization and entity should be concerned as well.

As a former journalist, it is concerning that people do not trust traditional, reliable sources of news; choosing instead to watch, listen and read news and information that only supports their point of view and makes them distrustful of mainstream information. As the owner of a public relations firm, it is concerning to know that no matter how transparent the information and messaging we develop for clients, some individuals and audiences will not believe it and may even publicly refute what is being said. And, as an executive committee member of a major sports franchise, it is nearly impossible to keep up with the wild west world of untrained citizen journalists churning out stories as they try to prove they have more reliable insider information than the organization itself.

The lack of trust is concerning on many levels. How can we possibly have healthy, positive relationships with individuals, companies and institutions we interact with if we don't trust them? On a broader scale, a lack of trust makes it hard to get people to work together to solve local, national and global challenges. It would be easy to simply throw up your hands and say that's just how things are, but I have not given up hope that there is a path back to trust and truth in information.

In *American Cyberspace: Trials and the Path to Trust*, Major General Mari K. Eder, U.S. Army (Ret.), provides valuable insights into the challenges of finding the truth in today's information apocalypse. But, more importantly, she provides clear direction on steps we can, and must take, to regain our sense of trust in the information we are hearing, reading, and viewing. This is essential reading for everyone.

Susan Finco

Susan Finco is the owner/president of Leonard & Finco Public Relations, Inc., an award-winning firm specializing in issues management; traditional, online, social and digital media; reputation management and crisis management. She is also a member of the Green Bay Packers seven-person Executive Committee, responsible for the strategic direction and major business decisions of the NFL franchise.

Sources:

https://knightfoundation.org/reports/indicators-of-news-media-trust/

https://www.pewresearch.org/politics/2019/07/22/trust-and-distrust-in-america/

◊ ◊ ◊

Our very survival depends on our ability to stay awake, to adjust to new ideas, to remain vigilant and to face the challenge of change.

—Martin Luther King Jr.

On a chilly April afternoon in Vaihingen, Germany, my troops gathered around as I had the honor and privilege of pinning on the U.S. Army Colonel collar devices (eagles) for then Lt. Col. Mari K Eder. I was the Chief of Public Affairs for U.S. European Command (EUCOM), a Navy Captain public relations specialist assigned to the unified command overseeing all U.S. military activity in 91 countries in Africa, Europe and the Middle East. She was the public affairs officer for the Marshall Center in Garmisch, Germany, for which I had administrative responsibility. Over our time working together in Europe, we formed a unique bond and I knew on the day of her pinning that wouldn't be the last of her promotions.

I left EUCOM and retired from the Navy not long after her promotion and joined the ranks of academe, where I've taught, advised and mentored thousands of undergraduate and graduate students in the art and science of public relations. General Eder continued her military career, reaching the rank of Major General. Among her assignments along the way was as the Deputy Chief of Army Public Affairs, where she championed the use of social media well before it became a mainstream means of official communication for the military.

Each chapter in *American Cyberspace* chronicles the destructive path this decay of trust is carving through our society. Maj. Gen. Eder has chosen the topics we most need to think about, asking questions that are both difficult but necessary if we are to find an antidote to what's happening in our society. Fake news, disinformation, misinformation, cancel culture and the corruption of civil discourse are epidemic today. Maj. Gen. Eder addresses these issues and more with unabashed candor and focus. At times, the questions she asks will make you uncomfortable. But it is the introspection required to answer these questions that will help us develop strategies for dealing with the impact of information overload and loss of trust in our lives.

Maj. Gen. Eder jump starts the ultimate resolution of this tangled web of challenges with clear insights and guidance for communicators in protecting and promoting organizational reputation. She addresses the best use of messages and channels to advance organizational goals. And she provides practical tips on how communicators can assist organizations in their communication with their stakeholders.

With more than 45 years of experience as a communicator and educator, I find *American Cyberspace* to be a much-needed addition to the body of knowledge in strategic communication. Now, more than ever, we need the insights, questions and arguments advanced in this book if we are to "stay in the game." It does our society no good to engage our "flight" instincts. We must stand and "fight" to defeat the challenges to societal norms so clearly outlined in this book. General Eder has given us a tremendous roadmap for that journey.

Bob "Pritch" Pritchard
APR, Fellow PRSA, Captain, U.S. Navy (Ret.)

Bob "Pritch" Pritchard is a member of the public relations faculty in the Gaylord College at the University of Oklahoma. He is the faculty adviser for the Stewart Harral chapter of the Public Relations Student Society of America at OU and for Lindsey + Asp, the nationally affiliated student-run advertising and public relations firm in Gaylord College.

◊ ◊ ◊

There is a language associated with trust especially now as words and visuals have never been more important in establishing a positive reputation with diverse and skeptical stakeholders. Mari K. Eder knows that words and pictures matter. She nails all the right phrases in this practical overview of the current issues facing communicators and their companies and organizations. More importantly, she offers clear advice to help guide communicators who are working to protect and promote their firm's reputation.

As program director for two professional communicators peer groups, I continually hear from corporate communicators about their daily dilemmas with the fast-paced and complicated multi-media landscape they manage, both internally and externally. They spend their days dealing with increasing expectations from many more constituencies and frequent crisis and issue management challenges, to name a few. As the country deals with COVID-19 and racial injustice, among so many other things, the general public is looking to businesses to step up in addressing workable solutions to these issues.

Digital platforms bombard people with information ranging from useful to dangerous. There is little distinction between work and home content consumption. The fuzzy line between home and work life has been growing for some time and accelerated greatly with the onset of COVID-19 and working from home. It's unlikely we'll ever be back to "normal."

The growing political and social polarization in the U.S. contributes to a lack of civil discourse that inhibits constructive change. We socialize with people like ourselves. We seek validation from online groups of people who think like we do. We rarely listen—really listen—to why someone might hold an opinion unlike ours. Most people are just waiting to talk.

At the root of all this discord is an essential lack of trust—in other people, institutions, corporations, and government.

This is the environment communicators face as they advise their executive leadership on messages, channels and audiences. It's often up the communications leaders to define the culture and reputation the organization wishes to cultivate. They act as the conscience of the organization. Communicators are the people who remind their leaders of the mission and vision that are critical to uphold and, in fact, to actively demonstrate through action, so the outside work knows what the company stands for. In recent months, communicators have had to become expert in pivoting—quickly changing strategy or tactics, but without ever changing the vision and mission of the firm.

Closely connected, and top-of-mind for communications leaders are the concepts of brand, perception and reputation. The tried and true definitions remain: Brand comprises the firm's values and purpose, expectations and promises, experiences and consequences. Reputation and public perception are how others see you, regardless of what you profess to stand for. Ideally, the brand influences reputation in a positive way.

With all this on communicators' plates, how do they make sense of the chaos? How can they recognize and plan for the next crisis? How do they help managers and executives communicate clearly and compassionately to audiences ranging from employees to customers to the general public? A great start is to learn from these essays based the distinguished career and deep expertise of General Eder. She's seen—and done—it all, as you'll see from her impressive biography. In a world of fake news, cyberspace scams and dizzyingly frequent messages, this is one authentic voice you can trust!

Janet M. Botz

Jan Botz is the former Dow Corning Company Director of European Communications in Brussels as well as the company's former global Chief Communications Officer. She later served as Vice President of Corporate Affairs and Communications at the University of Notre Dame. She established Botz Communications Consulting in 2012 and works as a program director for two communications councils of The Conference Board. Jan actively supports Public Broadcasting, serving as president of the PBS Wisconsin Friends Board.

◊ ◊ ◊

One primary characteristic of today's society is the growing public distrust in democratic institutions. Public trust in many traditional authorities, like political parties and the media, has hit rock bottom, causing them to fall from public grace. This trend is alarming because it leads to increasing cynicism towards the very fundamentals of democracy. Our national agendas and activities become influenced by the individual actions of people in other countries. For instance, many of the fake news websites that emerged during the 2016 U.S. presidential election have been traced to a small city in Macedonia where teenagers were producing sensationalist stories to earn cash from advertising. The 2016 election not only introduced new words, such as *fake news* and *alternative facts*, but also revived many Cold War archaisms, including *disinformation* and *kompromat*.

The internet used to be praised as a free and democratic environment, which enables open and frank discussion on key issues that affect society. This is no longer the case. Today, the internet still facilitates connections among people, but also makes them more vulnerable to fake news and conspiracy theories. Internet anonymity is a double-edge sword. On one hand, in situations of limited press freedom or suppressed public expression, securing anonymity of the source is a prerequisite for safety, privacy, and open discussion. Anonymity also provides a cover for whistle-blowing and investigative journalism. On the other hand, internet anonymity is easily abused by unrestrained, impulsive or manipulative behaviors such as trolling, harassment, and fraud.

News sources frequently report on falsehoods and inaccuracies they find to be newsworthy, thus unintentionally giving them more exposure. The analysis of the rise in far-right online activity mentions that it would be less significant if the mainstream media had not amplified its messaging. According to the study, the mainstream media was susceptible to manipulation from the far-right press due to a number of factors including: low public trust in media; a proclivity for sensationalism and novelty over newsworthiness; lack of resources for fact-checking and investigative reporting; and clickbait and corporate consolidation resulting in the replacement of local publications with hegemonic media brands.

Today, many scholars and practitioners raise concerns about the vulnerability of democratic societies to misinformation. The proliferation of media sources and the reemergence of more partisan news outlets has reinforced tribal divisions, while enhancing a climate where facts are no longer driving the debate and deliberation. The impact of deceptive online communication is significant, as falsehoods and fabricated stories do not only lead to public confusion, and political polarization, but also contribute to distrust in main institutions.

Communication scholarship is often evaluated on the extent that it generates new insights and produces new thinking. *American Cyberscape* helps scholars and practitioners navigate the murky waters of today's media environment. The notion of "information apocalypse" provides multiple opportunities for future research of the relatively new area. If we agree that internet communication has a dark side, then it is time that scholars start paying closer attention to various forms of online deception by which social actors seek to manipulate each other.

Under the conditions of trust deficit, there is a great demand for new voices to trust. With influence being distributed away from traditional institutions, new leaders are expected to become a vanguard of democracy. Eder's book plays a very important role here. In addition to offering a critique of our ailing information society, it also offers communication professionals hands-on recommendations on how to approach various challenges of the internet age.

Dr. Sergei Samolenko

Sergei A. Samoilenko, P.h. D. is a professor of communications at George Mason University and Co-founder of the university Laboratory for Character Assassination and Reputation Management (CARP).

PREFACE

The question is always the same: Why this topic? And why now? The answer is that the topic of truth and trust in communication has never been more important than now, due to the disintegration of the public's trust in institutions and the failure of the Fourth Estate to maintain standards of factual reporting and decency in social commentary. As a former journalist, public relations professional, and communications expert, I have long been concerned with the impact of culture on our public institutions. Over time, I've become convinced of the need to not just comment on what is happening in the world around us, both in physical and cyberspace, but to look at trends with a strategic focus and attempt to learn what lies ahead and most importantly, how to influence it.

In 2018, I wrote a wide-ranging essay on the topic of communications and culture, "The Information Apocalypse: Is Already Here." The essay was published by the U.S. Army War College on their new website, WarRoom. After that, I thought I was done with the topic. But as time went by, I continued to research, comment, and write more. The single article became a series. Then I began to speak on the topic— at St. Edwards University, Edinboro University of Pennsylvania, The Naval War College, The

CyberSecurity Summit, the International Association of Business Communicators (IABC), at the U.S. Military Academy, and more.

As the series grew longer, my research expanded accordingly. I have been truly gratified to see how a number of universities, associations, and institutions are now conducting their own research into communications topics that contribute to both media literacy and civics lessons for democratic institutions. These include but are not limited to: The University of Washington Center for an Informed Public, George Mason University's Research Lab for Character Assassination and Reputation Politics, and the U.S. Army War College's proposed Applied Communications Lab in its new academic facility's Decision Innovation Hub.

I am grateful to a number of people who have helped propel me along this journey, most notably, Dr. Tom Galvin at the U.S. Army War College, Prof. Sean Costigan at the George C. Marshall European Center for Security Studies, Dr. Michael Hannan at Edinboro, and Mr. Sebastian Warren at IABC.

One of my former bosses in Army communications was fond of quoting: "In this field, the fun never ends." That is the whole truth and nothing but.

Mari K. Eder
Maj. Gen., U.S. Army, Retired

INTRODUCTION

The very concept of objective truth is fading out of the world. Lies will pass into history.

—George Orwell

Today, what I call the "Information Apocalypse" is clearly evidenced by the unraveling of trust in American institutions. This phenomenon affects not only democratic institutions, their products and outputs, but also news, information and most critically, ideas and values. The decay of trust goes beyond institutions and also affects individuals, families and social norms. Personal data and identity are at risk. At the same time, the ever-expanding role of technology is not only omnipresent in our lives, it also acts as an accelerant, speeding up not only change but our abilities to keep pace and to control our responses.

Institutions: The Foundation of Democracy

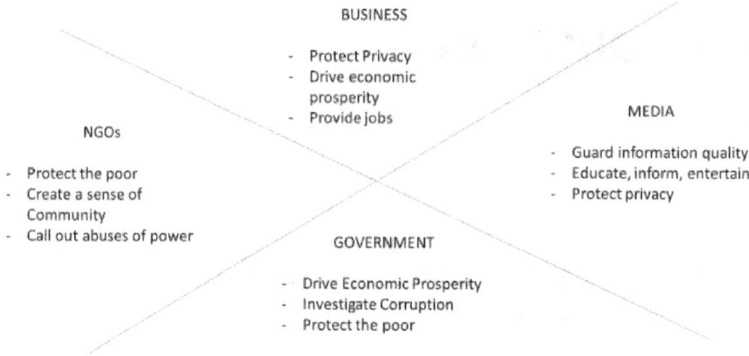

BUSINESS

- Protect Privacy
- Drive economic prosperity
- Provide jobs

MEDIA

- Guard information quality
- Educate, inform, entertain
- Protect privacy

NGOs

- Protect the poor
- Create a sense of Community
- Call out abuses of power

GOVERNMENT

- Drive Economic Prosperity
- Investigate Corruption
- Protect the poor

Over the past several years, skepticism has spiraled downwards to a sour layer of cynicism, outrage at civic life and the ongoing destruction of trust in systems and institutions. Everyone has been affected and infected. Ultimately, this has been personalized as private individual accounts are hacked and personal reputations are sullied, if not outright ruined. The results a forced introspection— does our very way of life, investment plans, credit usage, and even online spending habits contribute to vulnerability in both the physical and online world? Can news reporting any longer be counted on to warn us of the dangers?

The scourge of disinformation, misinformation and propaganda have taken up residence in the Fourth Estate. Absent the ability to fully rely on the mainstream media for truthful reporting and factual information, what organization or institution assumes the role of watchdog in a democracy? U.S. society is being flooded with information operations campaigns, psychological manipulation of response mechanisms through deep fakes and other outright lies, hostile hacks and takeovers of cyberspace organizations—by state and non-state actors. It is, to borrow a phrase, the perfect storm, now subsuming not just institutions but each of us personally.

Technology is understood to be an accelerant and amplifier that has resulted in the destructive path widening; speeding up not only the pace of change, but our abilities to understand and absorb the risks and impacts associated with emerging fields, such as telemedicine or artificial intelligence. The pace of innovation is sparking concern in every sector of society, from increasing fears about reductions in privacy and the ability to maintain one's identity, to the weaponization of information against populations, and the impartiality of the code behind artificial intelligence used in decision-making. How we address, manage, and control these immense challenges is becoming increasingly critical to our future roles in exercising the elements of national power, from diplomacy, to information, economics, and the use of armed forces in a military response to threats.

One institution in which public confidence is critical is the national intelligence community (the IC), comprised of seventeen difference organizations whose job it is to assess threats against the nation and citizens. The IC provides advice and recommendations for response and change, often based on incomplete information and sources with varying degrees of reliability. Previous missteps by intelligence agencies, ranging from drawing inappropriate conclusions from limited intelligence material, to insufficient or untimely cooperation across agencies, and even being susceptible to attacks, have resulted in grievous errors in national security decision-making. The IC has a number of challenges to meet in order to restore its credibility and the trust of elected officials, and yet it too is beset by efforts to politicize its work.

A critical examination of sources, reliable and unreliable, leads us back to the central role of the Fourth Estate in our changing world, where democratic institutions and professions are struggling to remain trustworthy and demonstrably ethical. The media plays a critical role as institutions become increasingly fragile and their inability to remain impartial and uncritical is becoming increasingly divisive.

As technology's intrusiveness into our personal lives continues to expand, the corrosive effects of a lack of personal privacy in the 21st century fast lane and the inherent clash of personal privacy with operational security is brought into a focus daily. The assaults on private information continue nonstop: from banks, credit card companies, poor safeguards on genetic testing, medical procedures, and insurance, among others. From the broadcasting Fitbit to spying voice assistants and even corrupted video doorbells, it's never been more difficult to protect one's private matters.

Leaders, both elected and appointed, play a crucial role in this uncertain world, where concern over privacy and protection from business exploitation of personal data, can impact future personal and business opportunities and society's battle with income inequality. The growing perception of unfairness in life, coupled with a perception of a growing lack of well-being, threaten democratic institutions whose brand image—and ultimately the trust we place in them as good stewards—is under fire.

From declining trust in institutions and organizations, we are witnessing declining faith in professions that adhere to established set of standards, medicine, journalism, and law, for example. This mistrust extends from the profession itself to suspicion of those who engage in it, particularly in the professional judgement of those engaged in its practice. The professions that hold sway in our lives—from the scientists whose calculations, research, and logic collides head on with emotion and superstition, often resulting in a multiplicity of interpretations that come with every new discovery. And when there is no common basis of agreement and all facts—indeed the very process that creates facts—are in question, the result is societal paralysis.

Our civilization is now sidelined by the pandemic, mired in illogic and a mess of confused responses and resurgences of cases, openings, closings—all formed by faltering institutions that are seemingly unable to advance. We can't even come to agree on a direction in which to move. On all fronts the future is in question. Our present state continues to be battered by a pandemic, volatile stock markets, double digit unemployment, and fears about the future and the state of democracy. Faith in the election process leads us down the road of questioning what impacts we can possibly have as individuals and is it worth it to take on the trolls, scammers, liars and others. Quarantine, then, may become a sort of national hibernation: Let's emerge when it is safe to breathe, to interact in socially acceptable and courteous ways, and engage in civic processes such as voting, when and where we can do so without fear of criticism. But the future waits for no one.

We must rebuild our innate sense of optimism for the future and consider that in a post Information Apocalypse world, it is possible for our digital lives to transform, supporting a positive transition to a new knowledge age, where privacy is protected, data is only shared through permission and truth shines through.

CHAPTER ONE

Trust, Lost

We are careening towards a future where the ability to distort reality shakes the foundations of democracy...

—Aviv Ovayda (@metaviv) via Twitter,
Feb. 12, 2018, 5:08PM

No, there aren't any formerly-human zombies. But there are bots and trolls, hackers, hijacked accounts, zombie networks, propaganda, and pretenders of all sorts. Lies and liars abound. How do CEOs, military leaders, clergy, and influencers—not to mention the people and organizations who support them—succeed in this toxic information environment and maintain the bond of trust with their communities? How can national security leaders continue to work with allies, recognize enemies, and respect ground truth? How do you level set? How do we engage a weary public on critical national security issues?

This is the real danger: that people become numbed by the daily dose of news disguised as outrage. From there, whether disappointed or disgusted, they may simply give up, stop challenging the army of falsehoods that confront them and disengage from public discourse entirely. Before that happens (and even after) how can we reach them? What can be done at the seat of government or by other institutions? What can we do as senior leaders? The answer in part lies in the ways and means of telling truth and preserving trust, even as the deluge of misleading and false information threatens civility and productive public discourse.

The Gallup management consulting company annually publishes the results of its poll on the most trusted institutions in the country. While trust in newspapers and broadcast media has fallen in recent years, other American institutions have fared even worse. Notably, the military retains its high standing (Shane 2019). In 2016, internet news—a new category—was rated fairly low, although recent privacy issues and manipulation of content may make that figure drop even lower. Today, the news media is the most distrusted institution in the world (Schudson 2019). Some of that distrust may be linked to the mainstream media's slide—whether witting or not—into sensationalism and a standard process of rushing to judgement in nearly every crisis.

	June 2018	June 2019	June 2020
	%	%	%
Newspapers	23	23	24
Public Schools	29	29	41
Banks	30	30	38
Organized Labor	26	29	31
U.S. Supreme Court	37	38	40
Criminal Justice System	22	24	24
Congress	11	11	13
Television News	20	18	18
Big Business	25	23	19
Small Business	67	68	75
Police	54	53	48
Church/Organized Religion	38	36	42
Military	74	73	72
Medical System	36	36	51
Presidency	37	38	39
News on the Internet	N/A	16	N/A
Large Tech Companies (new category)			32

Chart depicts the percentage of respondents who expressed a great deal of confidence in the institution listed.

Gallup

Unfortunately, this lack of adherence to focus and standards coincides with a substantial rise in propaganda (American Historical Association 1944). From its demise in the popular lexicon following World War II, propaganda was for decades confined to political advertising or enemy broadcasts and to many was easily recognizable (American Experience, no date). In the past twenty years propaganda slyly insinuated itself into the mainstream, promoting particular points of view and causes with limited factual information. Propaganda also became associated with celebrities and was the means by which they built their brand image, padded their resumes, and remediated or inflated their reputations.

There is a well-defined American cultural tendency to place public figures on a pedestal and revere them for their accomplishments. But the list of those who have been toppled from those pedestals since the 1990s is a long one, packed with prominent names hailing from almost every revered institution, including: the clergy (most prominently the Catholic Church abuse scandals), sports (e.g., doping of Russian Olympic athletes, The Houston Astros trash-can banging to announce pitcher's signs to their hitters, and professional football's deflate-gate controversy), political scandals including those of governors like Schwarzenegger, Spitzer, and the forced resignation of Secretary of Labor Alex Acosta when his plea deal with sex offender Jeffrey Epstein when he was a federal prosecutor in southern Florida was revealed to the public. Scandals from the world of entertainment are too numerous to mention. But one name today is more than worthy of mention: Harvey Weinstein's sexual assaults and rapes; his abuse of power over many years in Hollywood, served to galvanize and solidify the "MeToo" movement. Finally, the celebrity reporters and well-paid news anchors from the mainstream news media itself come crowding in.

As we witnessed a souring of ethics in public life, we also became inured to our loss of confidence in those reporting the news (Josephson 2005). Print and broadcast journalists have also become major celebrities, with fans and followings of their own. Eventually, many fell from grace, including the likes of Brian Williams, Charlie Rose, and Matt Lauer—victims of their own arrogance and false sense of invincibility.

Even entertainment-world celebrities, with their long history of nonconformist lifestyles, now fail to titillate in quite the same way with their affairs and excesses. The debasement of language in movies, television, and song seems to have reached rock bottom itself, until the next shocking revelation. Yet we seem unable to look away. Celebrities now influence more than personal choices in attire and cosmetics, they influence issues and even markets. Who could forget how in February 2018 Snapchat shares fell six percent in response to a tweet by celebrity Kylie Jenner who said, "Soooo does anyone else not open Snapchat anymore?" (Ungarino 2019)

The result of the toxic mix of American culture and apparent lack of standards is a slurry of political excess, celebrity indulgence and hubris. To some considerable extent this negative atmosphere existed prior to the election of the current president. Yet that event marked the beginning of a coarser new era and a significant change in social norms, inviting disquieting lows in civil discourse, the standards for public service, and promoting excesses in personal behavior. In one of the initial interviews regarding his book, published in the wake of his firing, former FBI Director James Comey stated, "We are living in a dangerous time in our country with a political environment where basic facts are disputed, fundamental truth is questioned, lying is normalized, and unethical behavior is ignored, excused, or rewarded" (Kakutani 2018).

Unfortunately, military leaders have not been absent from this stage. Sensationalized stories about the very public fall of former Generals David Petraeus (Tapper 2012), Michael Flynn (Corn 2020), and Stanley McChrystal (Rogin 2013) should come to mind. The bottom line is that the very public failures of public figures further erode our faith in the institutions they represent. We need our leaders, imperfect as they are, to live up to the standards they spent their careers espousing. We can't be expected to believe propaganda instead.

So we should ask: Can we believe much of the news we see and read? Or perhaps, if one accepts that there are potential issues with factual reporting in all news coverage, the larger question becomes then: Why aren't social media companies like Facebook regulated by government? Why aren't there ratings informing the public about veracity or reputation for impartiality?

Many restaurants in the U.S. are required to post their sanitation ratings, yet when we read a political piece on a social media site there is no warning that it isn't safe for human consumption. How do we know what we are getting? This is a singularly important question since most of us have the inherent tendency to seek out news that confirms beliefs we already hold (Stucky 2016). According to Irqa Noor, a neuroscience and linguistics student at Harvard University, this factor affects not only news consumption but the search for confirmation of religious beliefs, politics, and human resource concepts (Noor 2020).

Even the well-traveled Media Bias Chart by Vanessa Otero that purports to assess various media by its focus is subject to interpretation; many may disagree with its placement of news sources as liberal, conservative, or ultimately, even unreliable.

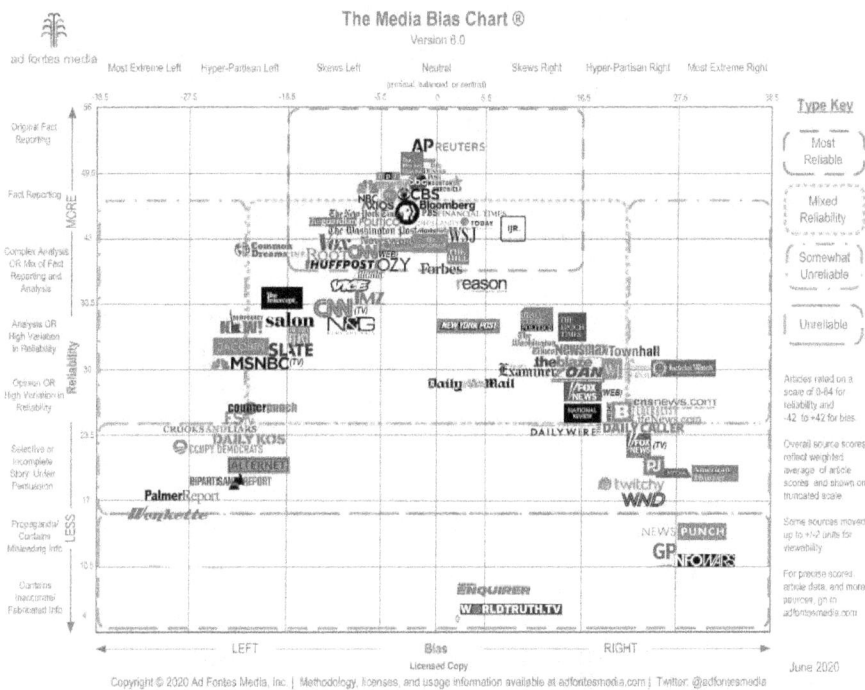

The Media Bias chart is provided via license from Ad Fontes Media.

While dangerous before, viral growth of these technologies and changes regarding expectations of behavior continues to grow at a pace matching that of the COVID-19 pandemic. The issue of Russian interference in the 2016 presidential election has alarmed members of both parties and magnified the issue of social media responsibility (Mueller 2019).

The result of this interference is undeniable. Following the conclusion of the U.S./Baltic Summit, the National Security Advisor, then-Lt. Gen. H.R. McMaster said, "Russia brazenly and implausibly denies its actions, and we have failed to impose sufficient costs. The Kremlin's confidence has grown, as its agents conduct their sustained campaigns to undermine our confidence in ourselves and in one another." (Atlantic Council 2018) Fears and unsubstantiated concerns about the 2020 presidential election continue unabated.

Worldwide, more than two and a half billion monthly users access Facebook (Mohsin 2020). This global phenomenon has flattened distance and engaged and networked people across the world in ways never seen before. With the explosive news that Cambridge Analytica had obtained data on 87 million users through a third party developer and that data was used in an attempt to influence voters in the presidential election, Facebook was forced to acknowledge its responsibility for protecting consumer data and its obvious failure to do so. As a media platform it is eclipsed in the U.S. only by YouTube, which is used as a source of news and entertainment by seventy-three percent of all adults and ninety-four percent of Millennials (Matsa 2018). With the explosive news that Cambridge Analytica had obtained data on 87 million users through a third party developer and that data was used in an attempt to influence voters in the presidential election, Facebook was forced to acknowledge its responsibility for protecting consumer data and its obvious failure to do so.

Naturally the question about regulation surfaces repeatedly in the political theater that constitutes a public Congressional hearing. But the questions remain: Should American lawmakers focus on regulating tech companies or permit them to continue to regulate themselves? Are social media companies the same as media companies? Facing lawmakers' questions, Facebook's CEO, Mark Zuckerburg promised a number of changes to how Facebook is managed in an effort to increase security, authorizing ads, and protecting privacy.

It is a different story entirely in the European Union. There the General Data Protection Regulation (GDPR), which came into effect in May 2018, maturing in its efforts to better protect privacy both through strictly regulating "Big Tech" data collection and enforcing consent. The GDPR has teeth and the regime has already begun to fine companies. At some point, the U.S. Congress may follow suit with an American version of such a law, but it is likely not to be a swift or easy change.

Delving deeply into issues of institutional trust, the Edelman Trust Barometer series looks beyond information manipulation by foreign governments, trolls, political operatives, or false flag organizations to the more primal concern that information can be used as a tool of warfare itself.

Generally speaking, global statistics on trust remain fairly static but by 2018, the Barometer revealed a staggering thirty-seven percent drop in trust in all institutions in the U.S. According to the Barometer, nearly sixty percent of adults said "I am not sure what is true and what is not," while over fifty-six percent said, "I do not know which politicians to trust," and forty-two percent said, "I do not know which businesses to trust." In 2017, the most widely shared fake news story was that former President Barack Obama had signed an executive order banning schoolchildren from reciting the pledge of allegiance. The story was read more than two million times. During the 2016 election cycle. By one account, the fake news had circulated nearly forty million times.

Perhaps even more worrisome is the portion of the Edelman trust study that reveals rising global concerns regarding fake news as a weapon (Quell 2020).

In 2020, just a few months since the beginning of the pandemic, over sixteen countries passed new laws criminalizing the distribution of fake news about COVID-19. Yet many countries had no need for new laws. There were already laws about the spread of false information on the books. China and India are two of the nations with the most stringent laws. Fake news has also disrupted elections in South Africa. In Singapore, the government is considering new laws to combat fake news while Germany now fines media companies for *failing to delete* fake news.

In the U.S. trust issues are broader than that of influence or manipulation but go to actual distortion. Artificial intelligence (AI) is a major new and emerging factor in this area, from the long-criticized alteration of news photographs and digital video editing, to computer enhanced or even created images, often not designated as such. Associated issues include AI used for decision-making in a military context, to direct drones, satellites or other weaponry, all when a certain set of criteria are met, and in conditions devoid of human oversight (Rouse 2020).

China is investing heavily in developing AI and its related technologies (Westerheide 2020). While AI has obvious and not-so-obvious business applications it can also be used in military operations. For example, future computers could direct swarms of bots or order satellites to attack other satellites and destroy them. Once programmed with a set of parameters for decision-making, it may be difficult to override those requirements. The danger here is complexity of code, biases that may lead to error, and speed of application. Could warfare speed up to the point that people are unable to keep up with the decision-making process, or in the language of military AI strategists, stay "in the loop?" Or is it that humans are "out of the loop" and are no longer even part of that process? We have a preview of these possibilities from our front row seats here are at the beginning of this change, as we watch the evolution of the driverless car and its decision-making capabilities. In some instances, the code or technology creates accidents where none would normally have occurred, but in the vast majority of cases humans cause errors in conjunction with these systems—not to mention outright mistakes.

Just as the manufacturers of driverless cars have to work hard to build consumer confidence, other businesses have to continue to strive to protect privacy and build prosperity. Edelman states that employees typically trust their employers to do the right thing, with a global confidence level of 72%. Government and non-government organizations (NGOs) have further to go in restoring trust.

The institution that is in last place is the Fourth Estate. In order to fulfill its self-described mission to educate, inform and entertain, the broad institution of media must do more to guard information quality, discipline itself and protect the privacy of consumers. According to a recent study by the Pew Research Center, nearly six-in-ten Americans (58%) say they would prefer the public's freedom to access and publish information online, including on social media, even if it means false information can also be published (Mitchell 2018). That tradeoff is pivotal, as it means individuals who reject regulation must somehow do a better job of recognizing false information. How can we, as individuals, detect and challenge mis- or disinformation? Individuals can:

- ✓ Ask questions. Ask for evidence. Ask yourself if it is a joke.

- ✓ Check sources. Is the URL for the story legitimate? (For example, abcnews.com and whitehouse.gov are legitimate sites. Abcnews.com.co and whitehouse.org are not.)

- ✓ Quotes by a public figure can also be fact-checked and traced back to an event or statement.

- ✓ Photographs can be examined by reverse searching the URL on Google.

- ✓ Pause for reflection before sharing.

As producers of information and official positions or commentary, business and government leaders should aim for consistency and transparency in communication. They should perform as role models of professionalism and always act in accordance with stated organizational values.

- ✓ They must respond early to lies, outrageous charges or fake news. Even if a commander doesn't have all of the correct information or the right answers, it is more important to be able to say, "We are going to find out. We will investigate and will tell you what we learn."

✓ Know when to pre-empt potential negative news. There are times when internal briefings to family members—telling them first about a potential deployment, or an extension—can go a long way to towards maintaining the bond of trust. Background sessions with local media that explain processes and procedures can also be helpful. "This is how casualty notifications work. Here is how the process for a court martial unfolds."

✓ As for fake news or alternative facts, call it out. No drama, no accusations, simply demand the facts.

There are numerous sites that bill themselves as fact checkers or scam debunkers but many are ineffective. But awareness is the most important step, and the one that should result in false information being exposed.

According to American University, there are currently ten reliable fact and bias-checking websites that are effective at examining news stories. Among others, these include Fact Check, a site sponsored by the Annenberg Public Policy Center at the University of Pennsylvania; Media Matters, a nonprofit research center; News Busters, a conservative site; Open Secrets, a website run by the Center for Responsive Politics; Politifact, an independent site that checks the claims made by elected officials; ProPublica, a nonprofit newsroom; Snopes, an independent non-partisan website; and the Washington Post Fact Checker.

In 2019, The University of Washington launched a new course titled, "Calling Bullshit: Data Reasoning in a Digital World." American University developed a game, "Factitious" to determine the credibility of a news story. Hopefully continuing innovation coupled with media scrutiny can give consumers the confidence to assess facts and recognize false narratives in news.

Over the past several years, the balance between misinformation and the tools to challenge it has continued to teeter. Upon reflection, historians might come to say perhaps this phenomenon might be better defined as a chaotic beginning to a new information age. At this juncture, it remains to be seen whether news, media, and Big Tech companies will take a positive role in righting the pillars of their institution or whether the Fourth Estate will suffer further self-immolation. Whether these changes will be for the good is something that all leaders can and should influence through vigilance and sustained engagement. There is no sitting out this out. It's here. No one will be left behind. There is only those who will be consumed or those who have learned how to navigate. We need to be the trusted navigators.

CHAPTER 2

This Time It's Personal

Information is a tool of warfare, and information is warfare itself.

—An Army Aphorism

Over the past several years two eminent institutions emerged from crisis both shamed and embarrassed. First is the U.S. Supreme Court. The American people watched it run headlong into the #MeToo movement and get sullied in the process. In recent televised confirmation hearings and news coverage jam-packed with references to sordid adolescent behavior, binge drinking, and sexual assault, the pedestal that Justice herself stands seemed cracked. It felt personal. The confirmation of Justice Kavanagh was simply the most recent part of a much larger story: the gradual erosion of the legitimacy of the Court, as evidenced by increasing public divisions over judicial appointments, decline of respect for the jury trial system, and increasing disapproval of the courts as a whole (Reddy 2019). Americans are continuing to lose faith not only in the Supreme Court's non-partisan nature but in the fundamental process of vetting potential justices (Willow Research 2019).

The other institution in crisis is the Catholic Church, still reeling from another round of charges and cover-ups. In August 2018, Pennsylvania Catholics were stunned following the release of a grand jury report that documented more than one thousand cases of clergy abuse across the state. A recent Pew survey found that many Catholics think these abuses are still ongoing (Gecewicz 2019).

Although Americans may be exhausted by the onslaught of institutional breakdowns and allegations of fake news, we may also fail to realize the danger in turning away from public life and engagement. We are being targeted—not just as a nation, but as individuals. The attacks are becoming more personal and trust in our public and democratic institutions is falling away. Mainstream Americans are turning away. In their place are vocal groups formerly on the fringes who may be most willing to engage, at least within their own tribe.

Most religious 'nones' say questioning religious teachings is an important reason they are unaffiliated

Among U.S. adults whose religious identity is atheist, agnostic or "nothing in particular," % who say _____ is a very important reason they are unaffiliated

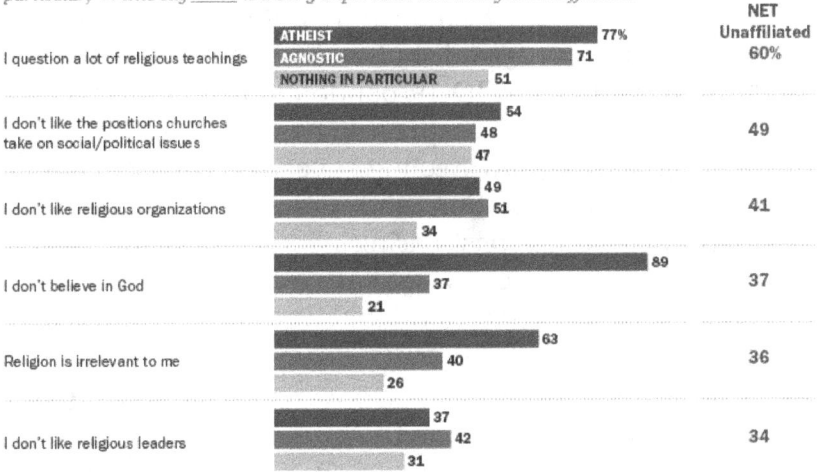

		NET Unaffiliated
I question a lot of religious teachings	ATHEIST **77%** / AGNOSTIC **71** / NOTHING IN PARTICULAR **51**	**60%**
I don't like the positions churches take on social/political issues	54 / 48 / 47	49
I don't like religious organizations	49 / 51 / 34	41
I don't believe in God	89 / 37 / 21	37
Religion is irrelevant to me	63 / 40 / 26	36
I don't like religious leaders	37 / 42 / 31	34

Source: Survey conducted Dec. 4-18, 2017, among U.S. adults.

PEW RESEARCH CENTER

A recent study, "Hidden Tribes: A Study of America's Polarized Landscape," discusses how minority groups on the far left and far right dominate the public discourse on major social and political issues. These hard-liners neither listen to nor respect each other. In the middle? The study calls these marginalized twenty-eight percent of Americans the *Exhausted Majority*.

Tuning Out...

The evidence of the disengagement of this moderate middle is growing, and this is a major societal problem. First, there was compassion fatigue, the name for a syndrome that originated as news organizations daily showed images of war, famine or stranded international refugees (Boyd 2011). While these heart-wrenching images and scenes initially spurred sympathy and donations, the long-term result was generally exhaustion, followed by apathy. Donor wallets snapped shut with a tired mouse click. Now with the COVID-19 pandemic well underway, quarantine fatigue is spreading (Marcus 2020). We have gone beyond feeling overwhelmed by the conflicting information and guidance we receive daily. People are past caring about #stayhome and many are tired or even combative of wearing masks, washing their hands and not knowing what will come next or when the crisis will end.

Not long ago, a Broadway play reflected the growing issue with regard to absorbing information. Titled "The Lifespan of a Fact," it starred Daniel Radcliffe as a magazine fact checker. The play examined the nature of facts and the fragility of truth, asserting that white lies undermine society's trust in itself. Politics takes center stage, entertainment for the disengaged. Now Broadway is dark, and entertainment sans social distancing is no longer an option for the disengaged or otherwise.

Engagement has its pluses and minuses. A recent Public Affairs Council pulse survey reveals that Americans who stay engaged with public issues tend to distrust national political and business news (Public Affairs Council 2020). Yet the news media remains the second most trusted source for news (forty-six percent), well below that of friends and family (seventy-one percent). But what if we cannot trust those closest to us? Now, mistrust of news and information gets personal.

Add this to numerous security breaches in personal data held by institutions from employers to banks, and we realize that we are under relentless attack all the time and from all sides. In a recent interview, Dan Coats, former Director of National Intelligence commented on the potential for a major cyberattack. He said unequivocally, "The lights are blinking red. Today, the digital infrastructure that serves this country is literally under attack." Bob Woodward's latest book, *Fear*, includes a similar warning about complacency: "I think people better wake up to the nature of the war on truth and its consequence."

While Adversaries Get Bolder

Warnings aside, there is increasing evidence that our adversaries are becoming bolder and more open with their attacks. A well-known quote by Albert Einstein states, "The world is becoming an increasingly dangerous place to live, not because of the people who are evil, but because of the people who don't do anything about it." It is in that context that I would like to us to consider what to do.

Physical and diplomatic threats take place daily on land, sea, space, and in the air. From Chinese ship incursions in the South China Sea to the growing Russian presence in the Arctic, to the persistent saber rattling of North Korea and Iran, threats to nations and individuals are continuing unabated. Yet the most pervasive and ever-present threat occurs in the information domain.

We have already felt its impact. The Federal Office of Personnel Management (OPM) suffered a massive breach a few years ago; among other key data points lost were the background investigation records of current, former, and prospective federal employees and contractors. All were stolen. In what has become a consistent refrain, major banks and even companies that purport to protect against identity theft have also been hacked. In 2018, Facebook promised to do a better job in protecting users' personal information. Yet, the company revealed in early October that a massive security breach had exposed the private data of twenty-nine million users, a security disaster that received only limited coverage. While companies and government didn't protect employee and user data, even worse was their lackluster approach to warning people about the effects of the breach. In an article for Slate Will Oremus reported, "It's a safe bet that those who took this data are not people you want 'pawing' through your personal life."

The hacks have continued to grow in scope and scale. Brazen takeovers of not only private information and celebrity social media accounts are now the norm. In July 2020, one attack targeted one hundred thirty high-profile Twitter accounts including those of Joe Biden, Bill Gates, Elon Musk and more. The attackers took over the accounts and asked followers to donate money, which would eventually make it to their Bitcoin wallet. Nearly as quickly, Twitter reacted by blocking the high-profile accounts, the attack stopped, and the perpetrators disappeared into the netherworld of the dark web. Writing in The Financial Times, editor Richard Waters said of the Great Twitter Attack, "Sometimes a security breach is so startling in its reach and audacity that it becomes a stark reminder of the precarious nature of or collective dependence on computer systems. The particular nature of the attack also serves as a commentary on the times."

Relentless personal attacks are increasing across all platforms and continue to reveal the fragility of social media and communications systems. People still fall prey to emails and other phishing scams that tell users to provide their passwords or click on a fake link. It is not just the internet that is at fault. Every form of communication is subject to misuse and abuse. Emails prey on others and snail mail can still ensnare the elderly with fake magazine subscription offers, Social Security scams and various cons. And let's not forget the phone; phishing calls use a wide variety of tactics to manipulate people. In one scam, a scam call frightens grandparents into sending money to help out a make-believe grandchild in trouble. As hurtful (and creative) as these social engineering schemes are, perhaps their worst effect is to foster a growing mistrust of information in any form.

Uncovering the truth can be time consuming and difficult. Columnist Geoffrey Fowler tried and found himself chasing down a number of rabbit holes to discover the falsehoods behind a popular Facebook video of a near plane crash. It had fourteen million views when Fowler discovered it. The faked video was sent to him by a friend. He talked to Facebook, to his friend and others in his investigation. The lesson: challenge something that seems too good (or strange) to be true.

It's a tall order but we must stop taking things at face value. Threats to the stable operation of that digital infrastructure, the internet backbone of global commerce and government relations, means that trust in communications institutions will probably continue to erode. This ongoing decay leaves the doors wide open for a plethora of other bad actors to engage.

Foreign Affairs devoted the full Sept/October 2018 issue to the information domain and its threats, titling the issue, "Word War Web." In *Like War, The Weaponization of Social Media*, P.W. Singer and Emerson T. Brooking argue that the internet is changing the nature of warfare. Twitter attacks can result in real-world casualties and disinformation, which in turn radically alters the discourse between governments and armies (Schake 2018). The information battlespace continues to blur both reality and perceptions of reality.

What the U.S. and Others are Doing

There is no greater indication of how serious this threat has become than the launch of police forces (not military) to fight it. On August 30, 2018, the FBI activated its new site dedicated to combating foreign influence. More individuals are being affected and the FBI is giving people both a chance to engage with law enforcement and providing them with the tools to fight personal attacks. The site grew out of efforts by the FBI's Foreign Influence Task Force (FITF). Established in 2017, the FITF was chartered to seek out, identify, and counter foreign influence campaigns targeting the U.S. The FBI continues to update the site, providing timely information and guidance on topics ranging from Russian election interference to character assassination.

While foreign influence campaigns have taken many forms over the years, the most widely known today are the attempts by adversaries, either governments acting covertly or non-state actors, to influence large segments of the U.S. population through false stories, advertising and sites designed to discredit institutions and political parties, policies, and public figures. A recent article, "Regulate to Liberate," explores the European Union's regulatory efforts to protect personal information. The GDPR, in effect since May 2018, asserts that personal data is a "fundamental right." This groundbreaking law continues to be closely studied by other organizations and governments as its applications continue to evolve. Certainly we will witness the courts take a role in the refinement of how the law is applied and interpreted. And other countries, including the U.S., will be watching to see what works and what doesn't.

We fail to realize the danger inherent in turning away from public life and engagement. In America, tech giants are working on internet security on their own terms. Twitter has been busy removing fake accounts from its site, even as more and different accounts pop up every day. Microsoft has been following suit, removing thousands of fake apps from its store. And Facebook? The social media giant says it is taking major steps to fight fake accounts linked to Iran and Russia. It is also serving as a watchdog for election meddling. But it has just been hacked again and continues to face criticism for its stance on hate speech.

Is relying on business to do the right thing enough? Frustratingly, it seems that efforts to counter online attacks are one step behind the next attempt to distort and destroy. Michelle Flourney and Michael Sulmeyer's article, "Battlefield Internet, A Plan for Securing Cyberspace," calls for the U.S. government to fundamentally rethink its approach to cybersecurity, and even consider the establishment of a new organization to focus exclusively on cybersecurity.

Dealing with macro-level threats is difficult enough, but threats to individuals are scary, damaging, and highly effective. Do you know anyone affected personally by a cyberattack? Rather, do you know anyone who has not been affected in some way? Phishing attacks go through Facebook and other social media sites in waves. Some fail; many still succeed. State and local governments, hospitals, movie studios, retail chains, law enforcement agencies, schools, state and local governments and private individuals are also targeted individually, randomly, and personally through network attacks, computer intrusions, ransomware and identity theft.

Just ask Kris Goldsmith, an Army veteran who is hunting fake Facebook accounts that target veterans. His work to expose and dismantle those sites has resulted in more than one hundred questionable pages, with millions of followers, being flagged to Facebook's managers. Many of these sites have Russian or Iranian ties. The FBI's site, created in concert with the Department of Homeland Security, provides a great deal of information for individuals, including a number of short videos providing tips and guidance for protecting computer networks and information on how to spot cybercrime. There is also a hotline for reporting internet crime. Their guidance for schoolchildren is set up by grade.

Down to Us

This is the bitter truth society must face: we cannot rely on our institutions to carry the fight alone. As individuals, we are all involved and have a role to play. There are no bystanders in this war. Individuals must rally to fight indifference and seek out the truth. We need to call out any information that is fake or misleading. This will not be easy. We have short attention spans. Even if we do pay attention to information designed to help us protect our personal information, there is more work to be done. We can limit our online profiles, safeguard social security numbers and birth dates, and never use Mom's maiden name as a password, but we still need more. We need tools and a *culture change* at a societal level that accepts and embraces the lessons of today's information age.

What more can be done to heighten awareness? Do we pay attention to news and educational programs designed to heighten awareness of the pitfalls of social media use? Perhaps not enough. Or the Department of Homeland Security (DHS) or FBI public information outreach campaigns? Perhaps not enough.

If we do not join in solving these problems and change our ways, we fulfill the warnings from the science fiction television series "The X-Files." Over twenty-five years ago Agent Mulder told his colleague, "Deep Throat said 'trust no one.' And that's hard, Scully. Suspecting everyone, everything, it wears you down. You even begin to doubt what you know is the truth."

Agent Mulder was right. There is more than credibility at stake. In the "Information Apocalypse," it is all about survival and hoarding our personal data as though it is the most precious thing we own. Because it is. While the series appeared to foretell a disenchantment with authority and reveal the beginnings of a fissure in trust in institutions, it also helped to create that same concept. Truth in "The X-Files" was an unreliable concept, one that bent to preconceptions and was molded by emotions. That viral idea is one that continues to spread.

CHAPTER 3

The War on Reality

The Terminator: "The Skynet Funding Bill is passed. The system goes online August 4th, 1997. Human decisions are removed from strategic defense. Skynet begins to learn at a geometric rate. It becomes self-aware at 2:14 a.m. Eastern time, August 29th. In a panic, they try to pull the plug."

Sarah Connor: "Skynet fights back."

—"Terminator II, Judgement Day," 1991

The war on reality is taking its toll. Daily, millions of people find there is somewhere else they would rather be than the here and now. Numerous companies, states, and non-state actors are all too happy to present a never-ending array of alternatives so as to benefit or profit from the sometimes desperate desire to escape the negatives of everyday life. This is the third facet of the Information Apocalypse, following the systematic breakdown of trust in societies and institutions, and the increasingly personal attacks on the character of those left behind in the wreckage. Alternative realities are becoming so attractive they do more than distract or entertain. Fictional new narratives can replace failed relationships, ease disappointments, build new worlds, and absorb us entirely. What we don't see are the lies and deceit that often mask their true nature.

Technology is the enabler for these alternate realities. In 2018 we witnessed a global drop in the trust index, trust in institutions faltered, and distrust in the news media increased. There was a growing fear that, overwhelmed by fake news, a coarsening of public life and the proliferation of not only theft, but attacks on reputation and standing, the majority of Americans would withdraw. It didn't happen. Instead, people began to latch onto narratives that they wanted desperately to believe, regardless of how outlandish or potentially untruthful. Yet for some the search for truth continued.

By 2019 engagement with news went up, more than twenty-two percent from the previous year (Edelman Trust Barometer 2019). One might expect to this to bode well for journalism—greater public involvement should mean the more thorough the fact checking and increased awareness of scams, falsehoods, and hacks. Indeed, some of the news is encouraging. More Americans are involved in grassroots movements from #MeToo, to #NeverAgain. Last fall's midterm elections voter turnout was huge, and independent news sources are rallying.

Spike in news engagement

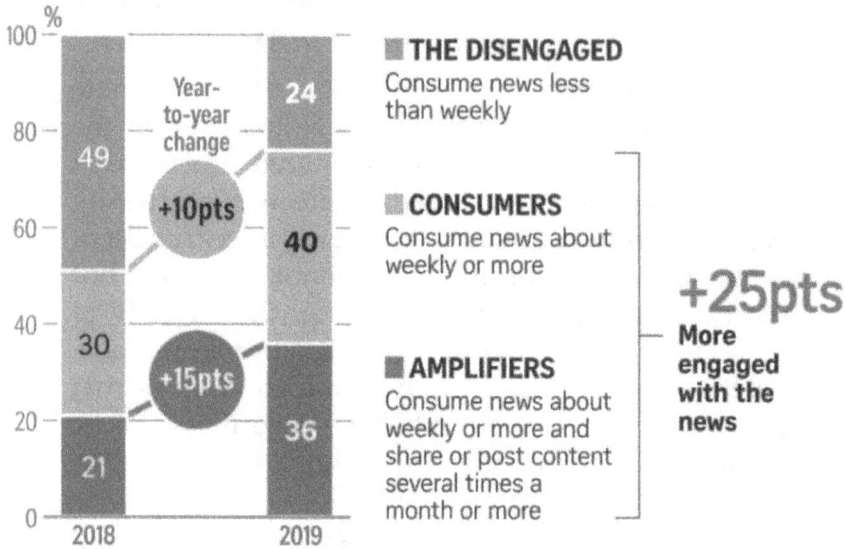

%

THE DISENGAGED
Consume news less than weekly

Year-to-year change

+10pts

CONSUMERS
Consume news about weekly or more

+15pts

AMPLIFIERS
Consume news about weekly or more and share or post content several times a month or more

+25pts
More engaged with the news

	2018	2019
Disengaged	49	24
Consumers	30	40
Amplifiers	21	36

Traditional media most trusted

■ Distrust ■ Neutral ■ Trust

Media source	Trust, in per cent, in each source for general news and information							
	2012	2013	2014	2015	2016	2017	2018	2019
Traditional media	75	68	70	62	63	62	70	71
Search engines	67	62	62	63	64	62	62	66
Online-only media	58	50	53	48	55	51	61	55
Owned media	49	45	48	43	45	41	41	48
Social media	55	45	46	47	45	41	45	46

73% Worry about false information or fake news being used as a weapon

Source: 2019 EDELMAN TRUST BAROMETER   STRAITS TIMES GRAPHICS

But populism and polarization are enabling our enemies in this war on reality. The enemies are not the technologies themselves—social media, gaming, AI, or even augmented/virtual/mixed reality. Rather, there are *actors* who misuse these technologies for their own purposes—to sow hate speech, to hack, to steal, to radicalize, incite violence, influence elections, or undermine democratic institutions. They, whoever *they* are, are both insidious and often invisible. Our enemies may, in some cases, also be those who are seemingly the most benign—the designers, psychologists and marketers who develop programs and games that seek to be so mindlessly entertaining and addictive they pervert perception, kill creativity and ambition, strangle decision-making, and lure people away from reality.

Big Tech/Big Problems

Are Big Tech firms our enemies, as some have suggested (Sherman 2017; Flood 2018)? Government efforts to regulate them and how they protect the data of individuals and corporations is growing. While many countries, particularly those in Europe, now focus on firms like Facebook and Google, America has thus far remained restrained in attempts to control these giants. Yet this battle is far from over, and it isn't just about gathering data in order to show ads. We are entering the phase where control of the information domain (and control of our choices) is the issue. From legislation to litigation, the response to these change agent platforms is standard, even if ineffective to date.

In the EU, the battle lines are becoming clearer. In February 2019, the British Parliament said that Facebook and other tech companies should be subjected to a code of ethics that would govern privacy protection issues and the laws for protection of competition. Germany has already ordered Facebook to curb its data collection efforts there.

France fined Google $57 million for breaching its new data privacy rules. French regulators said Google lacked transparency in gathering data to show targeted advertising and that user agreements were improperly executed. This is the first time a tech company had been fined since the EU's GDPR went into effect in May 2018. There are certain to be many more.

There is no clear agreement on the best ways to tackle technology's growth, protect privacy, and curb data collection. No one path has emerged as the clear winner in building support for a way forward, not by governments, or within Big Tech firms themselves. These companies are dealing with issues from data breaches, to hate speech, hacking, and manipulation by foreign entities and non-state actors. They are still reacting.

We are witnessing tech companies admit their naiveté. Their surprise that their creations can be used in ways they never programmed, designed, or conceived, is surprising. They, and we, have much to learn to protect ourselves better and still function — not just in society, but also in the military where the Common Access Card (CAC) card is our primary entry key. One day, and that day may not be far off, our enemies will find a way to stop its use or get inside themselves. Will we be ready?

There is a need for increased education, user awareness and a greater focus on individual as well as corporate responsibility. Tech companies should help users to determine the truthfulness of available information and be able to recognize sites and scams with ill intent, all without violating individual rights either to offer opinion or purpose. The bullies can be brought down, but the rules of engagement here still aren't clear.

Artificial Intelligence Meets Real Intelligence

Artificial intelligence has many definitions, but it broadly refers to code that has been programmed to make decisions or choices. It uses algorithms to make sense of data that is inputted, and while science fiction movies often depict how disaster results from the machines taking over the humans, the role of AI in society at large is still relatively small. Yet it is already pervasive and in use across a broad spectrum of business and government.

For instance, AI is used by numerous H.R. offices to review resumes for job hunters. Virtual assistants like Amelia, can even assist veterans with completing online forms for compensation claims, while others like Siri or Alexa are now household names. While employers may think the automated approach is more efficient and candidate-friendly, applicants and customers often find the approach impersonal and frustrating. AI's decision-making algorithms can be gamed and subverted by the savvy and sometimes even by chance. Isn't everyone doing it to get an edge? Find a job? Make a claim? AI's presence is often invisible, but its indifference to the individual human is becoming a significant influencer on behavior.

The pace of technology is worrying to many people. Edelman's 2020 Trust Barometer revealed that sixty-one percent of those surveyed believe the pace of change in technology is too fast. Even more strongly, sixty-six percent worry that technology will make it impossible for people to believe what they see or hear is real.

Robots may already be arriving, but the emphasis on the human aspect of leadership, personal interaction with the machine, and leader involvement are only going to grow as the pace and applications for AI increases. The arrival of 5G networks, with their promise of greater connectivity and the perils of possible new state-sponsored intelligence gathering, will only increase the speed of many applications and make other advances not only popular, but inevitable.

Gaming

Globally, gaming continues to grow in popularity and find new ways to connect with expanding audiences. In the introduction to her 2010 book, *Reality is Broken, Why Games Make Us Better and How They can Change the World,* Jane McGonigal began with a prescient quote from the economist Edward Castronova. She wrote, "Anyone who sees a hurricane coming should warn others. I see a hurricane coming."

She continues, "Over the next generation or two, ever larger numbers of people, hundreds of millions, will become immersed in virtual worlds and online games. While we are playing, things we used to do on the outside, in 'reality,' won't be happening in the same way. I think the twenty-first century will see a social cataclysm larger than that caused by cars, radio, and TV, combined."

McGonigal agrees with this analysis, asserting that gaming will continue to grow, stating that it feeds a critical human need for connection, and that games will become a way to collaborate and work for future change. That view of game appeal appears real, certainly fantasy worlds may seem like a terrifically attractive alternative to reality.

"Fortnite," the wildly popular first-person shooter game, had over two hundred million players by November 2018, with an average of 8.3 million players online at any one time. The game continues to break records, holding a virtual concert with an avatar for the electronic dancer-producer Marshmello in early February, 2019. The ten-minute concert attracted millions of viewers and the promotion racked up more than thirty million dollars in sales of virtual goods, from weapons to clothing to dance moves.

Is gaming the future virtual marketplace where people go to interact, play and do business? Game developers think so. In fact, Esports is expected to generate $345 million in revenue this year in North America, in addition to the more than half a billion dollars overseas. Large purses are generating new businesses, training hopeful Esports players for the national stage and building teams ready to compete in the international arena. The average salary for a player on a team in Riot Games' League of Legends North American League was over $300,000 in 2018.

Even U.S. Army Recruiting is now active in building an Esports team. "If we are going to be successful in recruiting, then we need to be where young people are—and they are operating in the digital world," Maj. Gen. Frank Muth, commanding general for U.S. Army Recruiting Command, said in a statement. Over sixty-five hundred soldiers applied to be considered for the team to date. Now fielded, the Army team is competing in a number of major tournaments, raising awareness and hopefully Gen Z interest in an Army career.

There are other avenues to access separate realities—virtual reality, which at this time requires specific equipment to use, such as a headset and goggles, to augmented reality, which can be used with a number of devices and shared. The augmented reality sandtable is already proving an inexpensive way to conduct a digitally interactive rock drill with multiple user, site, and participation options. The 3D battlespace map is easier to use, faster to set up and can improve and speed up decision-making.

It also makes actual war seem unreal. To what extent might gaming mask the reality of war, such that gamers inspired to serve by endless nights playing "Fortnite" or "Call of Duty" find themselves overwhelmed in a real shooting war? How will future soldiers deal with the uncertainties and fog of war that were not programmed into the games they were playing? How will they respond to real-life injuries or trauma?

Coming Attractions

The coming years promise more challenges for society as our enemies continue to exploit success and construct alternative realities that are even more attractive and lucrative than anything tangible. While the push to regulate tech firms and enforce corporate responsibility may increase, our enemies will be versatile and pursue private data through other means. For every hacker, bot, troll, false account, or influence campaign that is exposed and stopped, AI may generate dozens of new ones. The competition is heating up, alluring members of society whose trust in reality is broken beyond repair.

POTENTIAL REMEDIES		
	Reimagine systems	Societies can revise both tech arrangements and the structure of human institutions – including their composition, design, goals and processes.
	Reinvent tech	Things can change by reconfiguring hardware and software to improve their human-centered performance and by exploiting tools like artificial intelligence (AI), virtual reality (VR), augmented reality (AR) and mixed reality (MR).
	Regulate	Governments and/or industries should create reforms through agreement on standards, guidelines, codes of conduct, and passage of laws and rules.
	Redesign media literacy	Formally educate people of all ages about the impacts of digital life on well-being and the way tech systems function, as well as encourage appropriate, healthy uses.
	Recalibrate expectations	Human-technology coevolution comes at a price; digital life in the 2000s is no different. People must gradually evolve and adjust to these changes.
	Fated to fail	A share of respondents say all this may help somewhat, but – mostly due to human nature – it is unlikely that these responses will be effective enough.

PEW RESEARCH CENTER and ELON UNIVERSITY'S IMAGINING THE INTERNET CENTER

What to watch for in the ongoing war on reality:

✓ Increasing focus on regulation of tech firms and a major focus on corporate responsibility for safeguarding privacy and truth.

✓ Increased government and business efforts to protect privacy, stop hackers, find and force out bots and trolls, delete false accounts, and call out influence campaigns.

✓ New applications for AI and the growth of facial recognition software application.

We are struggling with the role of truth in the Wild West of reality, both physical and virtual. Governments and institutions are trying to find a way to regulate and thus hold onto traditional standards for objective truth and maintain flat standards for determining the rules of engagement. Our enemies don't care. They want to find a way into our programs, our bank accounts, classified plans, security systems, and ultimately into our heads. It's all about control.

Senior leaders must keep a finger on the pulse of change, use tech developments to develop force capabilities, capitalize on emerging opportunities to surge ahead, and build solutions for a more lethal force. Even as we look to forge the future, it is time to heighten our defenses.

Tighten the screws on your social media presence, double password protect your investments and bank accounts, fact check your trusted news sources and stay alert. Fasten your seat belts; it is going to be quite a ride. And there isn't even an app for that yet.

CHAPTER 4

Trusting Intelligence

American political elites feel very empowered to criticize the American intelligence community for not doing enough when they feel in danger, and as soon as we've made them feel safe again, they feel equally empowered to complain that we're doing too much.

—Michael Hayden

It's that midnight kind of anxiety, where you wake up—or you're already unable to sleep—and you feel that twist in your gut because you don't know what to believe anymore. Does Iran really have a weapons-grade nuclear capability? Are the intelligence estimates correct? Or the analysis, the understanding, or even the media reporting? Does North Korea have a viable nuclear weapons capability? Is Saudi Arabia an ally we need to maintain? Is there a viable opposition in Iran? What will Russia or China do next?

That these questions continue to nag at you is indicative of an even deeper concern. You lack confidence in the intelligence community to have all the answers, or even some of them. How can confidence be increased, skepticism satisfied, and measures of effectiveness realized?

Even as survey instruments such as the annual Edelman Trust Barometer purport to measure trust in institutions, and other polls attempt to convey respect for them, one institution that is not typically measured, either by Gallup or Pew, is the U.S. intelligence community, a federation of seventeen separate U.S. government intelligence organizations who conduct the assessments and make the recommendations that support foreign policy and national security policy decisions and their implementation. This is the singular community that Americans need to trust to feel secure in an increasingly complex, dynamic, and shifting global environment. But building that trust and confidence is challenging due to the classified nature of the business itself.

In July 2019, the Chicago Council on Global Affairs published the results of a poll, sponsored by the Texas National Security Network at the University of Texas at Austin. The poll aimed to shed light on Americans' perceptions of the intelligence community. While the baseline study was initially conducted in 2017, the second poll revealed a moderate increase in public confidence in the intelligence community. A strong majority, sixty-nine percent, regarded intelligence favorably—and vital to national security, despite a lack of understanding how the agencies operate, collaborate, are supervised and respond to oversight.

Effectiveness of the Intelligence Community

How effective do you think the intelligence community is in meeting the following responsibilities? (% very/somewhat effective)

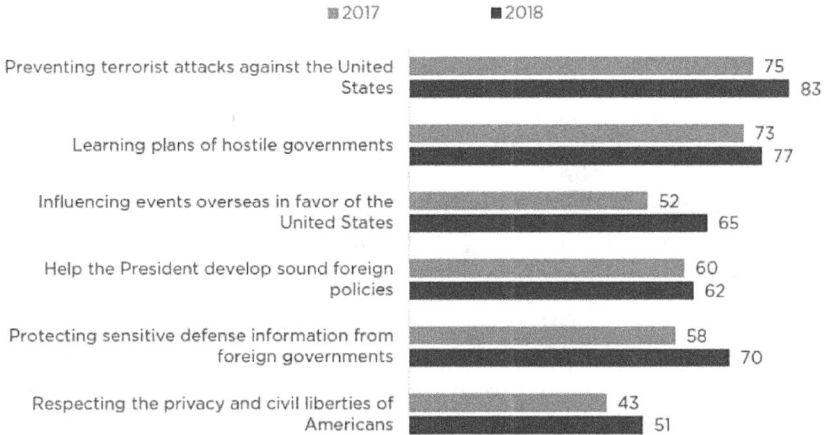

2017   2018

Responsibility	2017	2018
Preventing terrorist attacks against the United States	75	83
Learning plans of hostile governments	73	77
Influencing events overseas in favor of the United States	52	65
Help the President develop sound foreign policies	60	62
Protecting sensitive defense information from foreign governments	58	70
Respecting the privacy and civil liberties of Americans	43	51

July 24 - August 1, 2018 | n= 1000

While still serving as the Director of National Intelligence, James Clapper initiated a working group to develop ways and means to increase transparency and improve public trust in intelligence. The four principles in the council's charter were: to provide transparency about intelligence processes to improve public confidence, be proactive in making releasable information available, ensure intelligence professionals are responsible in the execution of their duties and to align resources with applicable laws and directives. Following the Snowden disclosures, Clapper later admitted, "Our adversaries...have learned a lot from our transparency. But in the end we think this transparency is worth the cost. Because if the American people don't understand what we are doing...why it's important...and how we're protecting their civil liberties and privacy, we'll lose their confidence...and that will affect our ability to perform our mission—which ultimately serves them."

The Vital Role of Intelligence

What is the business of this intelligence? Former Director of National Intelligence Clapper offers this simple definition: "*Intelligence* (author's emphasis) involves research, determination, persistence, patience, continuity, drawing inferences in the absence of complete information, and taking advantage of vulnerabilities and what you overhear in others' conversations, no matter how cryptic and jargon filled they are." But intelligence work, conducted in secret, must withstand deep scrutiny once the situation demands information be made public, or if information is leaked. That information must withstand the accompanying media frenzy, with pundits chiming in with their own spin on the intel and all the audience attention-grabbing *what if* questions.

Defining success of intelligence activities, particularly when it comes to countering foreign influence and disinformation, is more than difficult. Given the current operating environment, it is nearly impossible. The mountains of data generated every second are even harder to navigate than it may appear on the surface, particularly considering how easy it is for anyone to cover their tracks on the internet. As Americans, if we are concerned as a nation about countering foreign influence and disinformation, it is critical that we be able to recognize when our countermeasures are successful.

National security and policy professionals must rely on the intelligence community. But because it is clandestine by nature and shrouded in cultural baggage from movies and books about spies and espionage, it is a community the general public may not have equal confidence in.

Trust in the intelligence community's work is vital, more so than with other U.S. government agencies. After all, the intelligence community must act on limited and often incomplete information, occasionally propelled by hunches and innuendo. Intelligence analysts must also be on constant guard for deliberately misleading information, information operations and, increasingly, deepfakes.

Under Pressure

Media scrutiny comes regardless of operational success or failure. Perhaps the greatest example of success in recent years is that of the U.S. Special Operation Forces operation to kill Osama bin Laden. When President Obama announced to the nation on May 2, 2011, that U.S. special operations forces had killed the Al-Qaeda leader, there were celebrations in the streets of Washington. This was a major high point for U.S. intelligence and for weeks afterwards the public phone lines at the CIA rang with calls of congratulations and gratitude. It was a time of closure and one of faith that the intelligence community was effectively doing its job.

Conversely, perhaps the greatest example of the intelligence community's failure is that of Wikileaks' disclosures about U.S. intelligence operations, which broke faith not only with U.S citizens, but with

America's friends and allies. Wikileaks describes itself as a library of censored, classified, or otherwise restricted information. It claims to analyze that information and openly publish it. To date it has published over ten million documents and associated analyses. But the publication of leaks, particularly those that revealed information provided by then Specialist Bradley (now Chelsea) Manning, served to undermine trust in the intelligence community. While many of the leaks were embarrassing, they were not substantially damaging. But they definitely had a chilling effect on diplomatic and allied relationships. That effect is still ongoing, even as Julian Assange is now under arrest and pending a hearing relative to his extradition to the U.S. The U.S. intelligence community has identified Mr. Assange as an outlet for Russian propaganda.

So, just as the old saw about one man's terrorist can be defined as another's freedom fighter, one can also make a similar comparison between the self-proclaimed whistleblower and the spy. Leaked espionage may harm the national interest, destroy trust, expose sources and methods, and undermine alliances and relationships. Commenting on the fallout from Wikileaks to CNN's Fareed Zakaria, former CIA Director Mike Hayden said, "We've reminded our enemies how good and comprehensive we are at this. We will punish American businesses that have cooperated—under U.S. law and at the direction of a U.S. court. This is bound to be bad for them in terms of their international business. And then, finally, globally, a country or a source that might be thinking of cooperating with the United States, should have almost no confidence in our discretion or our ability to keep a secret. This is harmful."

Loss of trust is indeed harmful and difficult to rebuild. The Edelman Trust Barometer reveals a staggering drop in trust in institutions in the years since issues with Wikileaks first surfaced. Yet, while average citizens may display skepticism of intelligence activities and results, those in the foreign policy and national security communities retain the necessary faith and trust in the intelligence community. This is trust born of experience and professional respect, plus a confirmed understanding of how intelligence works in defense of the nation and in support of diplomacy, military missions and commerce. This confidence is forward-looking and based on both the working methods of the intelligence and about reputation and past performance. Decision-makers know that the craft of intelligence gathers, evaluates and objectively analyzes information and data, then disseminates conclusions for use, study or refutation. It's not perfect, and it shouldn't be politicized because it is necessary to the functioning of a modern state.

Metrics for Success

That decision-makers can rely on the quality of intelligence is influenced by four factors: the first is reliability. American intelligence organizations, from the CIA to the National Security Agency (NSA), are the best in the world. They are thorough, competent, and professional.

Second, intelligence professionals have integrity. They are honest about what they can and cannot reveal. And in their work, they are honest with their findings and recommendations to their superiors. All of our federal institutions are self-policing. The Defense Department and other federal agencies have a respected community of Inspectors General that continues to share best practices and information. The Council of Inspectors General on Integrity and Efficiency (CIGIE) is an independent executive branch organization that addresses issues of integrity, economy and effectiveness that transcend government agencies. Their training institute is focused on training Inspectors General (IGs) for work in an ever more uncertain and volatile environment.

Thirdly, intelligence must be dependable. Intelligence reports are a puzzle-piece story, a narrative often constructed from a long trail of varied sources, checked and cross checked, verified and vetted, carefully prepared and organized, often making sense as part of a larger whole construct or national expectation. In his recent autobiography, *Facts and Fears,* former Director of National Intelligence (DNI) James Clapper recounts the following story about General Colin Powell. As Chairman of the Joint Chiefs, he advised his intelligence briefers: "Tell me what you *know*. Tell me what you *don't know*. Then tell me what you *think*. Always distinguish which is which." (Emphasis by author.)

A renewed emphasis on Gen. Powell's approach to intelligence—a parsing of the knowns and the unknowns in supporting decision-making—is necessary to continue to build trust and faith in the intelligence community. The institutions and the professionals who conduct intelligence gathering, analysis, and prepare recommendations will continue to evolve, relying heavily on their colleagues, those seventeen organizations that coordinate, cooperate, and share intelligence, and contrast and compare analyses. It is critical that this work be accomplished clearly apart from the realm of political influence. A reputation for total objectivity is one that cannot be compromised. Products should be the result of professional efforts by an entire community with the goal of having a positive impact on the security of our nation.

One method could greatly enhance the intelligence community's reputation: a willingness to submit to a robust oversight or auditing process, led by Congress or an independent Special Inspector General. The weak Foreign Intelligence Surveillance Act (FISA) process could be brought to a greater level of legitimacy and efficiency. In August 2020 Congress approved another year's extension for the program, approving warrantless surveillance rules despite having found "widespread violations" of those rules. The FISA court's reviews are designed to protect the privacy of Americans when agencies conduct warrantless searches and then sift through mountains of data, not all of which is either relevant or gathered out of necessity. In many cases this has not occurred. The court rebuked the FBI for their flimsy justifications yet permitted the program to continue.

The Foreign Intelligence Surveillance Court was established by Congress in 1978. The Court entertains applications made by the United States government for approval of electronic surveillance, physical search, and certain other forms of investigative actions for foreign intelligence purposes. Certain congressional elements could carve out a greater role for oversight, buy-in, and understanding of other processes and programs, and their legal underpinnings, leading to greater support and inter-agency cooperation.

Issues, failures with intelligence gathering and analysis must be recognized and admitted upon discovery, even if only as part of a classified IG investigative report or to specific congressional committees, if too classified for general public disclosure. A true establishment narrative, one consistent with institutional identity and values, should always form the foundation of all IC communications. The IC must be able to consistently describe with purpose and clarity what programs, processes, and outcomes mean and how they meet specific national security goals and support U.S. interests. Through experience and repetition, the narrative will assume the reputation the IC community desires, one that can be trusted, relied upon, and respected.

Building Trust While Protecting Secrets

Finally, national security leaders must have faith that there is purpose to intelligence gathering requirements. Products should be the result of professional efforts by the entire community, with the goal of having a positive impact on national security. On the tenth anniversary of 9/11, Colin Powell said, "We have to be on guard that we don't spend so much time worrying about terrorism and guarding ourselves that we start to lose the essence of who we are as an open, freedom-loving people welcoming the rest of the world."

We may never know if there are unidentified flying objects (UFOs) following Navy ships. Even policy and national security leaders may not have the slightest inclination when North Korea will next conduct a missile test or on a whim decide to send the U.S. a "Christmas present." For that matter, regardless of our career, position or experience, national security leaders themselves may not know much about issues beyond their current highly specialized sphere of responsibility.

Some things we take on faith, and on trust, sometimes on instinct and our own experiences. But, when we depend on others to help us rest soundly at night, trust takes on a whole new meaning. We have to believe in the intelligence community as the nation's first line of defense against the hostile intentions and efforts of our enemies. If we cannot, no sleeping aid will help.

CHAPTER 5

The Fourth Estate

It is twice as hard to crush a half-truth as a whole lie.
—Austin O'Malley, *Keystones of Thought*

The Fourth Estate is in turmoil that has reached critical mass, yet traditional media continues to stagnate. Part of this state of affairs is the result of changing perceptions. The news media continues to be viewed as the most mistrusted institution in the world (Edelman 2020).

But now, the problems facing the media have gone to another level. There is a broad lack of trust in media as a profession. To take that one step further, there is considerable debate whether journalism even *is* a profession. The common frameworks of professions and professionalism legitimize this question. In an essay on "The System of Professions," Dr. Andrew Abbott discussed the evolution of a number of professions such as medicine and law. He termed journalism as more of a "permeable profession," characterized by frequent movement between journalism and public relations with considerable influence exercised by each side of that coin on the other.

After all, journalism is neither regulated nor credentialed. There are no annual training requirements, no enforceable ethical standards, no body of accepted practice similar to those of the legal bar or medical associations. There is no single code of ethics. Abbott noted that there is also no exclusion of those who lack journalism education, training, or adherence to any recognized code of ethics. While journalism may be comprised of people with the job to report, edit and write the news and to do so with a sense of purpose and ethics beyond that required by law or even society, this alone is not sufficient to meet the standards of a profession. Instead, what we loosely refer to as *the media* is a wide spectrum of individuals with varying capabilities, audiences, standards, and agendas. Has the quality of journalism suffered from this lack of professionalism, infused as it may be with public relations agendas, talking points and ideas? This omission has also limited the development of journalism as a vocation, perhaps making it less appealing as a career choice.

A further blurring of lines between objective reporting and influence is the degree to which the internet has democratized media. One could have made a stronger case for media professionalism back in the early days of print and broadcast media when numbers of outlets were limited and the perceptions of quality were far higher. Advertisers helped fuel these perceptions by providing dollars only to sources they viewed as particularly trustworthy or "professional." But this has changed due to the internet. Society's expectations of the media include "information quality, educating people on important issues, and helping inform good life decisions." Even

Mark Zuckerberg insisted for a time that Facebook was a platform, not a publisher. Yet, the American public has come to include social media in the category of "media." With social media being particularly vulnerable to fake news, this is a problem. Advertisers are pulling dollars away from traditional print and broadcast outlets to seek greater returns online. In turn, traditional outlets are trying to stay relevant.

People Define "Media" As Both Content and Platforms

What did you assume was meant by the phrase "media in general"?

Source: 2018 Edelman Trust Barometer. TRU_MED. In the above question, what did you assume was meant by the phrase "media in general"? General population. 28-country global total. Social is a net of TRU_MED10 and r12. Influencers is r5. Search is r7. Brands is a net of r10 and r11. Journalists is a net of r1 and r6. News Apps is r8

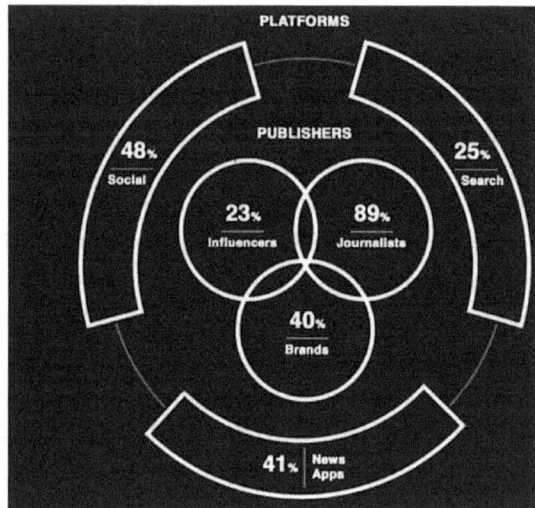

PLATFORMS
PUBLISHERS
48% Social
25% Search
23% Influencers
89% Journalists
40% Brands
41% News Apps

Thus far, that has been a losing battle. The news media today is vastly different than it was even two years ago. Newspapers have been declining in readership in the past twenty years, with weekday circulation dropping from sixty million in 1994, to thirty-five million by 2018. While local papers have seen the largest layoffs and drops in circulation, the increase in digital news access has affected urban papers as well. Advertising is also in sharp decline. In September 2019 The Washington Post abruptly ceased distribution of its free Express paper. The paper had an estimated readership of two hundred forty thousand. Twenty staff members were laid off. As the 2020s approach, one can envision a future without newspapers. Will local news sites fill the void? That may be logistically impossible. Local newspapers are in even deeper trouble. If communities continue to call for action like "defunding the police" and simultaneously lose any semblance of local news coverage, then who is held accountable in government, business, or any other institution that is the bulwark of a functioning society? The result would be the ultimate death of not only trust, but truth.

What of television news then? Can it step into the void? In general, television news is stagnant and, at least at the local and regional level, adhering to format and substance unchanged for decades. There is no discernable difference between local television news programs in 1989 and 2019—a half hour block with five or six stories, top story leads, with the last five minutes covering sports and weather. Substantive changes have been driven by cost. It is cheaper to rely on interviews than to actually cover events. And a recent Pew study showed that news stories were growing shorter while sports and weather began to expand. What little substance there is can be further diluted by mandatory banter among the anchors. This isn't news. It's old fashioned entertainment, where the anchors appear trustworthy and believable—and it's losing out to internet news.

It's also true the local target demographic is typically older and less interested in using the web as a resource. Yet, the rise of cable and the advent of the 24-hour news cycle has changed national and global news coverage. But the same stories tend to repeat and the moniker of "breaking news" is much like the cry "Wolf!" Often there is nothing new of note.

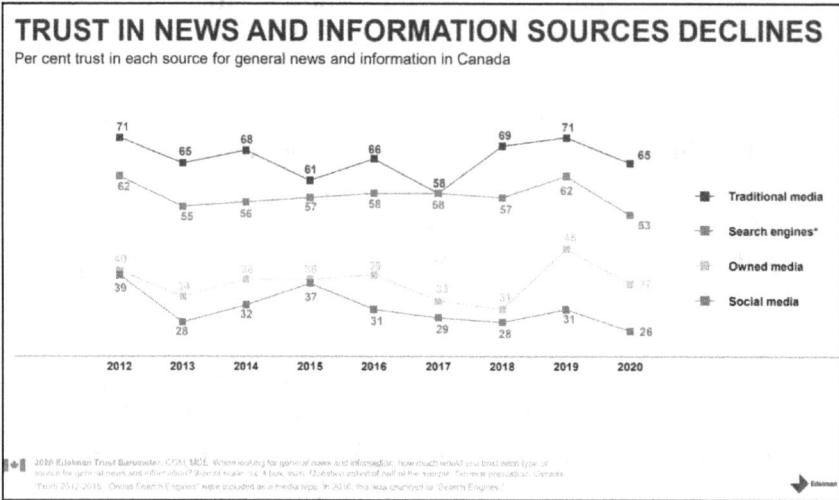

TRUST IN NEWS AND INFORMATION SOURCES DECLINES
Per cent trust in each source for general news and information in Canada

The notion of trust has been at the epicenter of our struggle. Here, "trust" is measured differently a bit differently, as the likeability and believability of individual journalists and new anchors comes into play. One could argue that if the news media is becoming less trustworthy, it could be because the older television news model is better at building trust. A recent study showed Americans trusted news that they could rely upon to be accurate, have the latest details, used valued experts as sources and provided information on sources as well as backup data.

Information Sources Influence Trust

We are witnessing a collision between two conflicting sources of information. On the one hand are the commercial sources that depend on the standard filter of traditional media and journalists. On the other hand, are the internet sources that are operating on a tenuous presumption that the online environment has flattened the world to the point that we can directly access ground truth and reality. Obviously the internet writ large has gifted us numerous conspiracy theories and ideas based on a false premise or faulty science, connecting previously disconnected conspiracy adherents. Think of the movement to prevent childhood inoculations for diseases like measles, or the fear generated in numerous communities by fraudulent science narratives that tied inoculations to autism.

As consumers of news and information we need a mechanism by which to judge: Is there a slant, agenda, or politics at work? Is an outlet decidedly biased right wing or left? Are we getting ground truth? People tend to seek out information that confirms what they already hold to be true; whether we may like it or not, algorithms analyze and amplify our choices. Internet platforms oblige accordingly with information displayed because "You liked this before." Psychologists call this "confirmation bias" — the tendency to seek out information that confirms or support beliefs already held.

Somewhere in between these news sources, an emerging model is creating space for itself, where crowdsourced and funded outlets appear to have the beginnings of a workable formula for building an audience and securing financial support. The approachability and the format itself, which can appear as decidedly anti-establishment journalism to some, is apparently appealing to many.

The Young Turks (TYT), a subscription news service on YouTube (and a paid app), is now the largest online progressive news and talk channel in the world, with 4.56 million subscribers. It features a number of additional outlets in its network, with top-rated shows featuring news, sports, politics and talk for the connected generation. Its reporters and anchors likewise appear unlike typical journalists.

Potential viewers are warned—these popular programs are aimed at "the 98% of people not in power." How does this channel pay for itself? A number of online fundraising campaigns have been highly successful, in direct contrast to the declining advertising model for traditional media.

The rise of new journalism models like TYT is leading the shift. Even Facebook, which long resisted hiring reporters and paying publishers, announced a new Facebook News Tab in October, 2019. In an interview Mark Zuckerburg said, "We get that the internet has been very disruptive to the news industry...This is an important moment in our relationship with the news industry and with journalism. We need to do a better job of supporting journalism." The news page launched in New York and plans are for it to become more widely available over the next year. The homepage will be managed by journalists, apart from editorial intervention by the company.

Twin Crises

The media's twin crises of relevance and professionalism requires both introspection and action. The first step is to come to grips with what purpose the profession serves—traditionally its role has been to inform, to educate, and also in some ways, to entertain. And because the profession serves society as a whole, access to media (access to the professional service) is a paramount concern. If indeed there is an audience that relies on traditional media because it is trusted more than social media, it becomes a professional matter to assure that access. It may be a bridge too far to regulate this unwieldly, widely-varied field that we call journalism or news reporting. Perhaps it would be more logical to define or divide by type and focus.

Certainly, we don't expect the same level of professionalism from gossip columnists or sports reporters as we do from global beat reporters who cover the Pentagon, State Department or the White House. But the dumbing down of journalism has led us to be better judges of what many of today's reporters are actually undertaking. The focus is often not on "telling a story," but to quickly rewrite a press release or regurgitate a spokesperson's talking points.

To Improve Trust

Obviously, there is more that media writ large can do from within to build trust and gain back its once-believing audience. The biggest effort journalism can undertake in serious news coverage is to show transparency. Who was interviewed for a particular story? What are other sources? Truth can no longer be taken at face value. Proof must be offered and willingly if insights are to be accepted. Broadcast journalists should disdain celebrity stories as news, resist cynicism and attitude in reporting, and stop repeating unfounded and salacious claims. Self-discipline in a profession notorious for its lack thereof can go a long way to rebuilding trust.

Journalists and editors alike can do several things to improve public trust in their work. They can easily increase transparency about their process for reporting and researching news. When serving as the subject of an interview, senior leaders should know they have a right to information about the structure of a story so that they can understand the reporter's focus, agenda, and how the reporter, as someone outside government, is attempting to make sense of what is happening within. For example, a recent Washington Post Fact Checker story on false and misleading video ended with the comment, "Our work on this project was funded by a grant from Google News Initiative/YouTube. Readers can find the full description of the categories with additional examples at The Washington Post's manipulated video fact-checker site."

Media literacy is paramount. While many schools in the U.S. have developed programs to teach media literacy, there is no national data on the efficacy of these programs. Aside from these programs, many adults have no idea where to begin to achieve this literacy. Senior leaders in particular need a sophisticated understanding of media processes and products. They need to be able to define what constitutes "fake news," know how to recognize "sponsored" content, and understand how social media platforms attempt to influence subscribers and the public at large.

At all levels, consumers need to understand how these challenges are driven by political and cultural agendas, and accelerated by the internet's global voices, are changing communication standards. Leaders must be determined to resist how the media time and space constraints tend to dumb down complex issues and multifaceted topics. Let's fight for context and detail, insist on transparency and continue to seek out engagement with media and with audiences of all kinds.

The Nieman Labs journalism predictions for 2020 draw on insights from a number of journalists. According to Colleen Shalby, a reporter with the Los Angeles Times, "We've been taught not to be the story, or divert from our priorities to inform the public and protect the truth. But if we want to continue to reestablish trust with our audiences and re-enforce our industry, now's the time to teach."

CHAPTER 6

Paranoia and Privacy

Every border you cross, purchase you make, call you dial, friendship you keep, site you visit…is in the hands of a system whose reach is unlimited but whose safeguards are not.

—Edward Snowden

Surveillance has been sneaking up on us for the past several years—a problem with Facebook stalking here, a hacking blip there, and even an issue with deployed soldiers being tracked through their fitness devices by unknown actors while out for a run on the Forward Operating Base (FOB) in a deployed location. We've been picking at the edges of the problem of protecting individual privacy, but we haven't fixed it, much less confronted it directly.

Apps as Trackers and Informers

The Department of Defense (DoD) kicked off the new year in 2020 with a ban on the Chinese-made app

TikTok. The app exploded in popularity in 2019, with its short video format and appeal to Gen Z users. The Army even used it as a recruiting tool, along with Instagram and other social media outlets, as part of its new recruiting strategy—the "Warriors Wanted" campaign. The approach focused on 30-second videos and memes as being more effective than television commercials at audience penetration.

Action shots of military hardware are the type of video subjects that go viral. That happened as planned on TikTok with scenes of paratroopers jumping out of helicopters or aircraft doing fancy maneuvers. What else did those videos reveal while showcasing weapons platforms, new training or advanced technological capabilities? In November 2019, Senator Chuck Schumer urged the Army to assess potential risks because the seemingly innocent app was busily vacuuming up user data, including Internet Protocol (IP) addresses, metadata, and other potentially sensitive information. Schumer said he was especially concerned about Chinese laws requiring domestic companies "to support and cooperate with intelligence work controlled by the Chinese Communist Party."

That resulted in the recent instruction to U.S. soldiers to avoid the app. The instruction also urged Army personnel and their families to uninstall the app from personal devices, phones, and tablets. This is not the first time a specific platform has been targeted for avoidance. In 2016 the DoD advised service members to avoid Pokemon Go because the app tracked the location of users and some users ventured into restricted areas on military installations while playing.

Smart watches and fitness trackers have also come under scrutiny. GPS-based geolocation features present in devices like smart phones, Apple watches, and other devices, including fitness trackers, were banned for service personnel while deployed. The DoD found that fitness trackers in particular posed a threat because they could reveal the location and travel patterns of soldiers when deployed.

Alexa listens and records. Doorbells watch and can share with neighbors and law enforcement what they learn. What does that data do when it is sent back to the company who programmed its harvest? It is more than shopping habits consumers need to be concerned about. It is the collection, sale to third parties, and use of personal data for surveillance, manipulation, and information operations.

A little over half of smart speaker owners are concerned about how much personal data their device collects

% of smart speaker owners who say they are ___ concerned about how much data their smart speaker collects about them

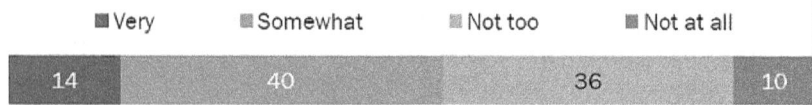

■ Very	■ Somewhat	■ Not too	■ Not at all
14	40	36	10

Source: Survey of U.S. adults conducted June 3-17, 2019.

PEW RESEARCH CENTER

The seemingly benign family genealogy sites that offer kits for individuals to trace their heritage through DNA research have come under fire. In 2018 the Golden State Killer was nabbed using information from an ancestry site that linked his DNA to a crime scene. The ethical and legal ramifications of the fact that personal kits inadvertently provided "evidence" to law enforcement continues. Whether companies sold the data they collected or researchers were able to access it through publicly accessible sites, the problem continues to be that private data released inadvertently could have major impacts—from paternity testing, to discovery of existing health issues, or even a genetic predisposition to cancer or other serious afflictions.

DoD officials were alerted to the issue last fall when it was brought to their attention that direct to consumer (DTC) testing companies, like Ancestry and 23andMe were offering discounts to service members to take the DNA test. Both companies stated that they were diligent in protecting user data and did not target service members. The DoD memo published Dec 20, 2019 states, "Exposing sensitive genetic information to outside parties poses personal and operational risks to Service members."

The means with which personal data can be harvested from unassuming individuals has grown exponentially over the past few years. It isn't just the technology that poses a threat, like fitness devices and entertainment apps, or even social media sites where individuals unwittingly disclose personal data in chatting or dating situations. It is the genetic fingerprint, whether exposed through family DNA testing or a retina scan that could potentially be used for mass surveillance or to identify individuals involved in a covert action, or lead to their discovery when entering a foreign country under a different name. The double-edged sword could be used effectively by law enforcement and military personnel but be weaponized by foreign governments or non-state actors for nefarious purposes.

Carrying Spies

Millions of people carry "spies in their pockets." The big question is: How do individuals protect themselves, particularly when the most innocent of cool devices, toys that connect to the internet, even simple downloaded gaming apps or maps all can potentially reveal each day's browsing history, reading choices, travel plans, destinations, and even health decisions. Preventing the data dump may be inconvenient. But it also may be necessary.

Surviving Surveillance

There is growing public awareness about some basic security measures:

- ✓ Don't accept the default settings on a communications device.

- ✓ Use complex passwords and two forms of authentication.

- ✓ Don't reveal personal information such as birth dates, hometown, or phone numbers on social media sites.

- ✓ Keep checking for potential breeches.

- ✓ Switch off all of the location monitoring access buttons on personal smartphones.

✓ Change the default setting from "Always on," to "Ask next time."

These steps may not be enough. It requires constant monitoring to keep on top of potential issues. Where to go for help? Will an app that stores passwords actually keep those passwords safe? The truth is that the majority of people don't have the time or the expertise to defend themselves from this nefarious offshoot—the Surveillance Apocalypse.

Protection of the Law

All consumers need to pay attention to changes in the law concerning data protection and privacy. The U.S. continues to move towards a broad privacy law like the European Union's GDPR, but it appears to be difficult at this time to gain consensus on how to move forward. Action to date has taken the form mostly of regulatory fines and a focus on tech regulation and anti-trust action.

States are taking action. In January 2020 the California Consumer Privacy Act (CCPA) took effect. It directs companies to tell consumers what data they have collected about them, and on request, to stop selling it. This requires companies to be cognizant about their data collection practices and storage. New York and Washington are considering similar laws and globally, India is also looking at a similar law while the U.K. will be developing its own privacy protections in the wake of Brexit.

The threat continues to grow. While changes are coming in laws and law enforcement, individual efforts only go so far. Perhaps consumers need something along the lines of a privacy advisor, performing the role of shield, teacher, and supporter in much the same way a financial advisor does. The new privacy laws in California and the potential for more in other states mean that there is an opportunity for start-up businesses to provide those health checks.

There is a growing number of apps designed to safely store passwords, delete old texts, or social media posts and photos, such as Lifelock, Private Photo Vault, and hundreds more. Jumbo is one new app that purports to "Help build an internet we can all trust. We believe that your data belongs to you, which is why we fight so hard to give you the tools you need to protect your personal data and privacy. Download today and take back your privacy!" New apps are arriving daily. Some increase parental controls. Others block participation in certain social media settings. Whether it's access, deletion, or simply opting out, changes continue to arrive with a downpour of options.

While it is impossible to predict where this will go, we do know that protecting ourselves is a lifelong project. We need to ensure that our technology is sufficiently advanced to protect us operationally and that we are educated to the point to be able to take measures personally to protect ourselves and our families.

Once upon a time, classes teaching "life skills" to high school students including such popular topics as checkbook management and the basics of auto mechanics were thought to be innovative. Media literacy and privacy protection may just be the next block of instruction we need to develop.

CHAPTER 7

Competence and Ethics

People grant their trust based on two distinct considerations: competence and ethical behavior.

—Richard Edelman

CEO, Edelman Communications

Just days before the Davos International Economic Forum in January2020, Edelman Communications released the results of their 20th annual global survey on trust. The results showed trust in institutions at record lows, what CEO Richard Edelman termed a "paradox." Despite generally good economic news over the past several years, skepticism and mistrust have been growing. This lack of trust is now deeply embedded in all democratic institutions.

While competence can basically be described as the ability to get the job done—or in military terms, completing a task to standard—performance alone isn't enough. For leadership to be trusted and respected, leaders must be seen not only as high performing, they must also be ethical—exhibiting behaviors characterized by honesty, fairness, and respect for others. Ethics, as the 2020 Survey revealed, is seen as three times more important than competence. Integrity, values, and reliability have never been more critical to society.

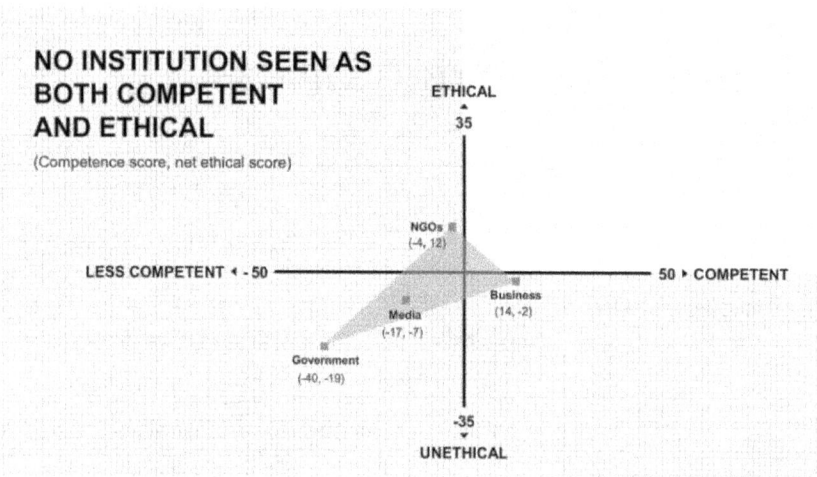

NO INSTITUTION SEEN AS BOTH COMPETENT AND ETHICAL

(Competence score, net ethical score)

ETHICAL
35

NGOs
(-4, 12)

LESS COMPETENT ◄ - 50 ———————————————— 50 ► COMPETENT

Business
(14, -2)

Media
(-17, -7)

Government
(-40, -19)

-35
UNETHICAL

Still, the general survey results were surprising. Competence has been analyzed in surveys for decades. Why did trust hit rock bottom in 2020? The paradox was in the metrics. In January 2020, just as the report was released, the global economy was strong. Unemployment was at near record lows. Even violent crime was down. Over the past twenty years nearly a billion people around the world have risen out of poverty. Yet the gap between the haves and have-nots has been widening. Fears about the future were rampant and respondents noted their unease with both democracy and capitalism. More than fifty-seven percent of survey respondents said that government serves the interests of the few.

Hopes and Fears

"The essential truth is: people are scared," Edelman commented. "Their fears are overcoming their hopes."

Like telltale tremors before a major earthquake, these results were revealed just before the coronavirus pandemic wiped out what little trust in democratic institutions remained. Beyond the immediate horrors of the virus' invisible spread, death tolls, and isolation, the coronavirus pandemic exposed deep fissures in the veneer of civilization. The growing inequity in wealth distribution was laid bare and the glaring disparity in access to education, availability of health care, and opportunity for advancement was evident across class and racial lines. Long lines at food distribution centers showed just how many Americans were living close to the edge. Elected leadership failed to establish a consistent approach to dealing with the virus across the nation and trust continued to plummet. There is still limited confidence in the competence of business leaders and in the ethical behavior of NGOs, but faith in *both* the competence and ethics of elected leaders and government is seriously lacking.

Early on, media reports highlighted the sacrifices of health workers and those providing essential services, from grocery store clerks to truckers. Concurrently, military service was recognized anew as an expression of selflessness and personal courage. It was risky to enlist. And it wasn't just doctors and nurses working with ill patients who were at greater risk of infection, servicemembers in uniform were also at risk, regardless of their daily routines. The necessities of military life—close living quarters and missions that render social distancing impossible—

exposed more risk not only to operational security but mission effectiveness. The result was an increased focus on leaders who could and would rise to the occasion—meet mission while protecting their subordinates.

CEOs and leaders at all levels reacted to the changed operating conditions they faced. When the virus struck the aircraft carrier Theodore Roosevelt, her captain wrote his chain of command, "We are not at war. Sailors do not need to die. If we do not act now, we are failing to properly take care of our most trusted asset—our sailors." He was swiftly relieved of command. Next, the acting Secretary of the Navy resigned.

News reports stated the captain had placed sailor's lives ahead of his career. That was the consensus of his crew on the ship. What did this situation reveal about military leadership, competence and ethics— by both the officer and the Navy leadership? The incident is still under administrative review and a medical investigation is ongoing. The answers to this question will be explored fully over time. One thing is certain. This leader definitely had the full trust and respect of his sailors. They thought his actions both competent and ethical. The case studies on this situation (and undoubtedly others yet to come) will continue long after a vaccine has been implemented, the raging virus at last brought to heel, and a new normal has been re-established.

In his recent book *A Time to Build*, Yuval Levin discusses how recommitting to institutions can revitalize the American dream. He said trust is based on "the belief that institutions have an internal ethic that makes their members more trustworthy."

One institution that has long been mistrusted for an apparent lack of that internal ethic is the police.

Following the videotaped murder of George Floyd at the hands of a police officer on the street in Minneapolis, the long simmering anger over treatment by police, the courts, and governments exploded in communities across the country. National Guard troops were deployed—many uncomfortably—to counter protesters. The Black Lives Matter movement became a rallying cry for social and cultural change across the U.S. and in cities around the world. The pace of change, real change began to accelerate, potentially opening the door to a renewal of trust in government institutions.

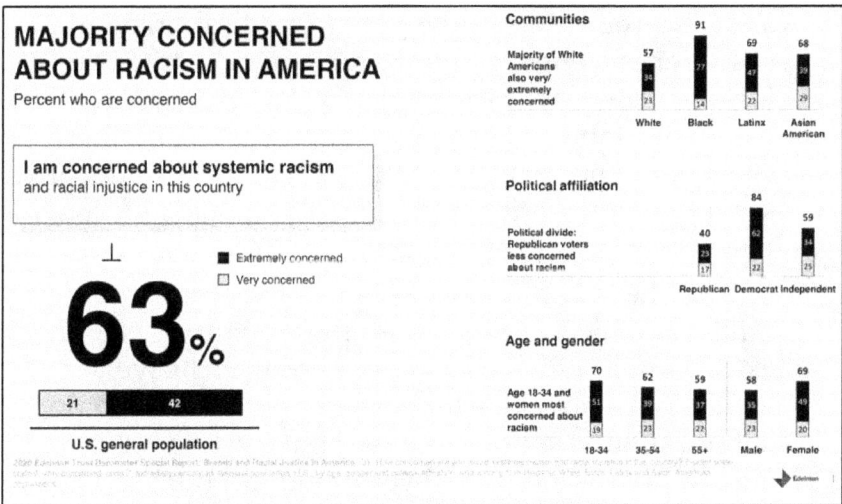

MAJORITY CONCERNED ABOUT RACISM IN AMERICA
Percent who are concerned

I am concerned about systemic racism and racial injustice in this country

■ Extremely concerned
□ Very concerned

63%

21 | 42
U.S. general population

Communities
Majority of White Americans also very/extremely concerned
White 57 | Black 91 | Latinx 69 | Asian American 68

Political affiliation
Political divide: Republican voters less concerned about racism
Republican 40 | Democrat 84 | Independent 59

Age and gender
Age 18-34 and women most concerned about racism
18-34 70 | 35-54 62 | 55+ 59 | Male 58 | Female 69

But for that to happen, structural racism and police violence had to be not only opposed but broadly acknowledged, changed, and in some police departments, "defunded"—then given a hard restart. In the past several months, years of transformational change has occurred. More is needed. But the role of leadership in this era is even more necessary at the national level—by elected officials, and by the military. At this inflection point, trust is fragile.

Chairman of the Joint Chiefs, General Mark Milley recently apologized for taking part in a photo op with the President after White House protesters were dispersed using tear gas and rubber bullets. "I should not have been there." he said, aware that his presence gave the appearance of military involvement in domestic politics. He later repeated those comments during graduation ceremonies at the National Defense University.

First he addressed the competence of the American military, noting the world order established following World War II is now under stress. He said, "We are, indeed, leading the Joint Force in dynamic and uncertain times. The recent medical crisis has cost over one hundred thousand American lives, and it has stressed our health system, our economy, and the social fabric of our communities. All of these challenges and many more will exist in the national security framework under which you, each of you, will operate as senior officers."

Then he talked about his personal values and the military ethic. "And while the military sets an example for civil society through our inclusiveness, we too have not come far enough. We all need to do better....Our responsibility as military leaders is to ensure that each and every one of our service members is treated fairly, with dignity and respect, and each of them is given equal opportunity to excel."

Through the honest admission of a mistake and a recounting of how he learned from it, plus his values-driven commitment to the Constitution, Gen. Milley righted the American military's focus from its recent and seemingly dangerous tilt away from adherence to the military ethic and values. If citizens and soldiers cannot find that balance of ethical behavior, conscience, and competence in their national leaders, they must be able to find it clearly visible in their leadership within the DoD. This is critical not only internally to the institution, but to the nation's relationships with its allies and international partners. Leaders at all levels must be particularly aware of the necessity to adhere to core values of fairness, impartiality, leading change and standing firm in the face of not only political influence but also the potential to refuse orders that go against the military ethic and the rule of law.

It is critical that those elected or selected to lead in uncertain times display not only competence but the highest levels of ethical behavior. In the military, an institution built on the bedrock of trust, it is imperative every soldier have the utmost confidence in the competence and ethical compass of their leaders. In uncertain times, particularly as other democratic institutions may appear unsteady, uniformed leaders, and those companies with effective leadership at the top will continue to lead the way.

Percent who say that CEOs should **take the lead on change** rather than waiting for government to impose it

76% ⬆ +11pts

Percent who agree CEOs can create positive change in:

Equal pay	65
Prejudice and discrimination	64
Training for the jobs of tomorrow	64
The environment	56
Personal data	55
Sexual harassment	47
Fake news	37

Edelman's 2019 survey found an eleven percent increase from the previous year in employee expectations that employers take the lead on creating positive change in society, a trend that continued in

2020 and has grown further as the result of new challenges including the global pandemic, calls for social justice and the need for leadership on the most pressing of social issues, ranging from immigration to the environment.

One example of leadership in this arena occurred in the summer of 2019 when, in the wake of the Parkland school shooting, the CEO of Dick's Sporting Goods announced the store would cease selling assault style rifles. By October the company had destroyed more than five million dollars' worth of weapons. CEO Edward Stack estimated that pulling the rifles from store shelves cost his company nearly two hundred fifty million dollars in lost business. When asked about the losses, he replied, "You know what? If we really think these things should be off the street, we need to destroy them."

While some company leadership teams may independently seek to improve their corporate responsibility standing, in some organizations, employees actively press their leadership to act. In April 2019, more than five thousand Amazon workers published an open letter to their CEO and board advocating the company take a stronger stand on climate change. Amazon had already committed to lowering its carbon footprint but the workers asked for an unwavering commitment to a number of specific goals.

This type of social activism continues to push important decisions up to CEOs—covering a range of issues that employees do not see being addressed by government or other institutions.

Even as institutions begin to rebuild in the wake of the pandemic, we can reasonably expect that business and in government, one institution—the U.S. military—will continue to lead the way in terms of developing trust, maintaining shared values and an ethical posture, and in growing leaders. The apocalypse may be upon us, but the drive for social and cultural change is showing the way for institutions to lead as we move forward.

CHAPTER 8

Trust and Science

Scientific knowledge is human knowledge and scientists are human beings. They are not gods, and science is not infallible. Yet, the general public often thinks of scientific claims as certain truths. They think that if something is not certain, it is not scientific and if it is not scientific, then any other non-scientific view is its equal. This misconception seems to be, at least in part, behind the general lack of understanding about the nature of scientific theories.

—*The Skeptic's Dictionary*

Declining trust eats away at the foundations of our institutions, our professions, respect for the pronouncements of professionals and the value of expertise. What remains is little but the validation of the concepts and approach of science, its logic, empirical processes, discovery, and the assemblage of facts that lead to facts, findings, and results. All of this has long been questioned as science has butted heads with institutions, professions, and individuals with bias, preconceptions, and opposing and perhaps contradictory, strong beliefs. What was once considered infallible is now being questioned at every turn. Yet American confidence in scientists to perform on behalf of the public interest remains high.

Americans' confidence that scientists act in the public interest is up since 2016

% of U.S. adults who say they have a great deal or fair amount of confidence in each of the following groups to act in the best interests of the public

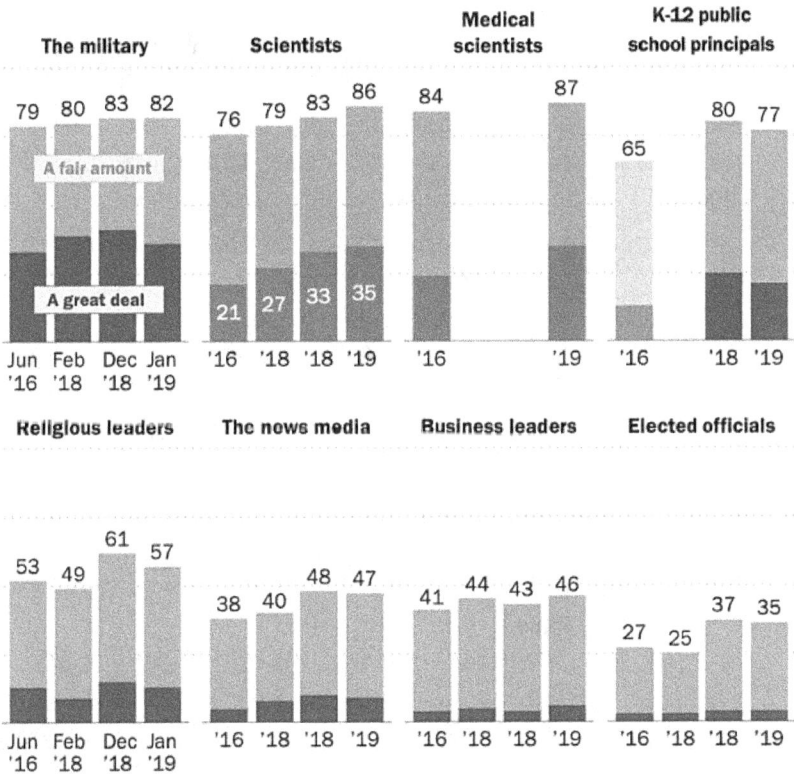

The military: Jun '16: 79, Feb '18: 80, Dec '18: 83, Jan '19: 82 (A fair amount / A great deal)

Scientists: '16: 76 (21), '18: 79 (27), '18: 83 (33), '19: 86 (35)

Medical scientists: '16: 84, '19: 87

K-12 public school principals: '16: 65, '18: 80, '19: 77

Religious leaders: Jun '16: 53, Feb '18: 49, Dec '18: 61, Jan '19: 57

The news media: '16: 38, '18: 40, '18: 48, '19: 47

Business leaders: '16: 41, '18: 44, '18: 43, '19: 46

Elected officials: '16: 27, '18: 25, '18: 37, '19: 35

Note: In 2016, question asked about confidence in K-12 public school principals and superintendents. Respondents were randomly assigned to rate either their confidence in "scientists" or "medical scientists" in 2016 and 2019. Respondents who gave other responses or who did not give an answer are not shown.
Source: Survey conducted Jan 7-21, 2019.
"Trust and Mistrust in Americans' Views of Scientific Experts"

PEW RESEARCH CENTER

Scientific Theory

Scientists are not infallible. And those who claim their results to be flawless will undoubtedly find themselves either ridiculed or exploited or both. The dichotomy between science and emotional beliefs is not new. The veracity of science has been discussed, argued, and fought over since the claim that the earth was round ran headlong into those who believed it, no—*knew it*—to be flat. From Adam and Eve to the theory of evolution, scientific inquiry and thought has been pitted against institutions and their preferred explanations—whether the church, the state, business, or media.

To be clear, the theory of evolution is not a "theory" in typical scientific terms. A theory typically is denoted as such because of a high degree of certainty that it is correct. A theory can be subjected to testing. Because it cannot be verified, scientists incorrectly refer to evolution as a theory just as they refer to Einstein's views on gravity as a theory.

Most recently we have seen several major conflicts in the arena of science versus anti-science. On one hand scientists claim, with evidence, that vaccines prevent disease. The opposition is led by the vociferous anti-vaxxer movement—who claim vaccines can cause disease, not prevent it and cite dubious internet sources as proof of every side effect and disease known to humanity, ranging from autism to cancer.

But science is not without fault. There were cases where flawed drugs designed for one purpose often resulted in unforeseen and sometimes tragic results. Thalidomide was marketed in the 1950s as a sedative and was favorably looked at as an alternative to barbiturates. But it was quickly criticized for causing nerve damage in some older patients. Then the tragedy of birth defects appeared. Today the drug is still in limited and controlled use, for therapeutic effects in treating leprosy, lupus and certain cancers. The science of the drug's development was sound. It was the safety protocols that were lacking and the broad application that was not only faulty, but negligent.

How Science Prevails

Science prevails when multiple independent studies, conducted in an impartial, empirical and repeatable manner reach the same conclusion. One must then balance the facts versus "alternative" facts and question whether delusion or mysticism is preferred. Even if delusion appears more attractive and science less comfortable, there is always the advantage to choosing the latter. This is true even if the science-based answer requires implementation of a course of action that is neither easy nor assured.

A different phenomenon is at work when science, even proven, is still perceived as questionable. That is partially the result of interpretation. We may not understand the science behind the creation of a virus vaccine, for example, the chemical formula that creates a soft drink, or the specialized engineering design for a new fighter jet, but people can and do question how that science is interpreted and communicated. It is the middleman, the administrator, the manager, the marketer, who make claims and seeks profitable application of scientific achievements and inventions, perhaps at times to the detriment of the end user.

Science versus the non-believers is a massive and wide-ranging topic. It goes beyond medicine, technology, and invention to our perceptions of the natural world. For example, there are polar opposites in this realm who disagree on virtually all aspects of climate change.

The non-believers refuse to believe that global warming is occurring. The believers fear for the future. One side cites the need for fossil fuels as reason to continue drilling, fracking, cutting and taking natural resources from the earth on a global scale—oil, gas, coal, trees, water. The believers want conservation, thoughtful awareness of consumption, a reduction in materialism, increased regulation of gas emissions, and more. It is a conflict destined to result in gridlock with neither side able or willing to compromise.

With the arrival of the global pandemic, something unusual happened. Quarantine meant people were staying at home. Flights were cancelled. Once jammed commuter lanes on highways were empty. The tourist trade tanked. True, that sizeable portions of the economy ground to a halt, inflicting terrific pain and loss on hundreds of thousands of families. But there were tiny silver linings in the dark clouds of spreading disease. As traffic congestion cleared, power plants slowed, and manufacturing began to lessen, satellite imagery showed airborne pollution beginning to plummet. In northern Italy, levels of nitrogen dioxide fell rapidly, by forty percent. Air quality improved across Great Britain, to the relief of asthma sufferers. In Venice, the canals grew clear without the ceaseless churn of tourist boats. Residents could see fish swimming there once again.

The pandemic certainly didn't provide the answer to environmental issues, but the planet took advantage of people staying home and began to heal. If anything, this effect was a massive, if inadvertent, experiment in how the environment could be improved through making more conscious and thoughtful choices about burning fossil fuels. The planet has issued a plea for change. Will our institutions—government, NGOs, the business community, and the media be able to act on this opportunity? Or will the press favor a return to "normal"—the push and pull of gridlock and the relentless march towards the disasters that worsening climate change will wreak?

Over the past twenty years Gallup polls have shown the numbers of Americans concerned about the environment have continued to rise. In 2020, about sixty percent now realize environmental quality is only poor to fair, continues to worsen, and that the government is doing little to help. More than seventy percent now favor increased restriction for vehicle emissions and power plants and a focus on developing new clean-energy alternatives.

Communicating with Transparency

Voices advocating change are becoming increasingly loud. Along with the planet's call for change, we have also received a sharp warning. Disasters, ranging from melting glaciers to rising seas, and severe health challenges such as COVID-19 will increase. Without a vaccine and compliance with recommended guidelines, the pandemic continues to rage. Having arrived on the scene, just as the Edelman 2020 study claimed that trust in all institutions was at an all-time low, the virus arrived when the bonds of credibility in every institution were strained. False narratives and misinformation continue to spread, with outlandish claims such as: the virus will disappear on its own, a vaccine would be used to implant microchips in patients, or that some individuals who took part in vaccine trials have died. This is a typical pattern for propaganda or disinformation—led by whispers of conspiracies and mistrust of government, of experts and scientists.

Dr. Anthony Fauci, Director of the National Institute of Allergy and Infectious Diseases commented recently on how the unbridled acrimony of public discourse affects science. He said, "There's so much extremism in things right now, it makes it very difficult. Whenever you want to be completely transparent about science and what it means, you have people who almost take that as an affront against them."

Yet, plain speaking is necessary. It is the only antidote—ongoing factual communication and public engagement. The developmental process, the science behind the research, the uses and access to data, it all has to be transparent. According to branding and communications expert New York University Professor Jacqueline Strayer, an essential rule of effective (and credible) crisis response is always *transparency*. The institution that will lead in this arena is business, picking up where the others are lacking in action or refusing to lead.

We can expect that the business sector will lead forward. A Edelman special report on brand trust released in June 2020, reveals that now, more than ever, people trust brands to solve problems and advocate for change. Brand trust has risen along with the issue of individual vulnerability—health, privacy, and financial stability. Over eighty-one percent of those surveyed said they trust reliable brands to lead the way ahead for society. This is true for both consumers and for private sector employees and companies with a focus on corporate responsibility. There are high expectations for business to act, and a great measure of reliance on CEOs to do the right thing. Consumers are continuing to demonstrate they respond to a brand's ethical stand on social issues.

2019 EARNED BRAND:
PEOPLE ARE BUYING BASED ON A BRAND'S STAND

Percent who are buying on belief

Belief-driven buyers:

- choose
- switch
- avoid
- boycott

a brand based on its stand on societal issues

Buying on belief now the new normal

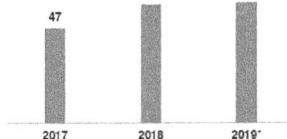

2017	2018	2019*
47	59	60

Black Americans most likely to buy based on a brand's stand

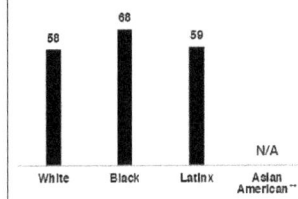

White	Black	Latinx	Asian American**
58	68	59	N/A

2018 Edelman Earned Brand. Belief-driven buying segments. U.S. Belief-driven buyers choose, switch, avoid or boycott a brand based on its stand on societal issues.
*2019 Edelman Trust Barometer Special Report: In Brands We Trust? Mobile Survey. Belief-driven buying segments. U.S. and among Non-Hispanic White, Black, and Latinx populations.
**Asian American population has too low of a base size to report out.

Edelman

Another Edelman report primarily focused on response to the pandemic, noted that in eight out of ten countries surveyed, individuals responded that, "My employer is better prepared than my government."

The basics of communication have much to teach about how to restore trust in institutions, professions, and people. Continuous and open communication is key—with a goal of transparency—plain talk, zero blame, no politicization, and even a smattering of humility can go a long way to restoring even a little trust.

It is the breath of fresh air we need.

CHAPTER 9

Civics Lessons

So long as we have enough people in this country willing to fight for their rights, we'll be called a democracy.

—American Civil Liberties Union (ACLU) founder
Roger Baldwin

We've established, and polls and surveys have confirmed, that Americans' trust in democratic institutions is at rock bottom. Given that fact, how can we safely approach open elections with any semblance of confidence in the voting process? Fears are multiplying: whispers of election interference by Russia or China or both, voter fraud via mail-in ballots, voter registration restrictions, and pandemic-driven fears over in-person voting. It is enough to make regular citizens throw up their hands and stay home, head under the blanket until the whole sordid mess is over and done with. Let someone else deal with it all.

That shouldn't happen. We need a deeper assessment of what is actually occurring, not more hand-wringing about what might happen. First, how do foreign governments interfere with elections? In 2016 Russia interfered in the U.S. presidential election in a sweeping and systemic manner, promoting one party's candidate over another, messaging to affect Congressional races, and sowing general discord and distrust in America's democratic election process via social media. The main features of their campaign included hacking campaign servers, internet trolls and bots repeating false narratives, intrusion into state election systems, messaging designed to limit African-American voter registration, targeting voting blocs and institutions and widespread hacking of internal voting management systems. The intelligence community reacted strongly and numerous investigations ensued. The lingering bitter taste in the public's mouth meant that future elections would likely be targeted again, leaving voters both nervous and afraid.

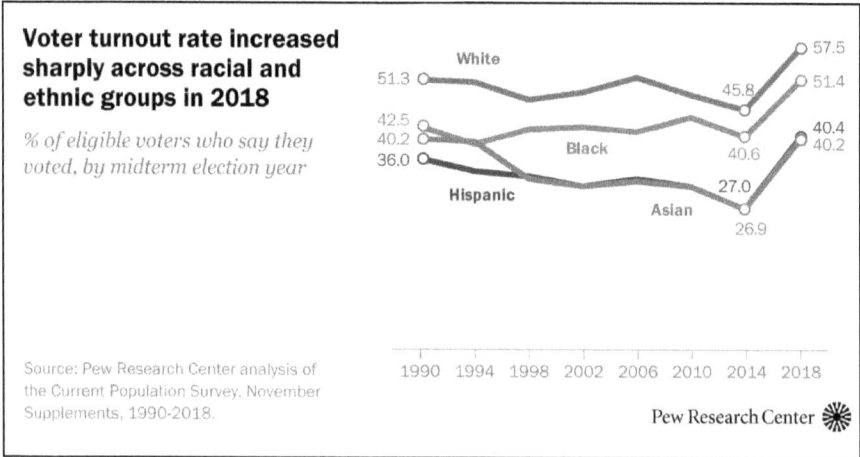

Voter turnout rate increased sharply across racial and ethnic groups in 2018

% of eligible voters who say they voted, by midterm election year

White: 51.3 ... 45.8 ... 57.5, 51.4
42.5, 40.2, 36.0
Black
Hispanic
Asian: 40.6 ... 40.4, 40.2
27.0
26.9

1990 1994 1998 2002 2006 2010 2014 2018

Source: Pew Research Center analysis of the Current Population Survey, November Supplements, 1990-2018.

Pew Research Center

In November 2019, in an unprecedented act, a joint statement was released by the DNI, FBI, DOJ, DoD,

DHS, NSA and the Cybersecurity and Infrastructure Security Agency (CISA) on ensuring security for the 2020 elections. The agencies all vowed increased vigilance and unprecedented information sharing. The release stated, "Our adversaries want to undermine our democratic institutions, influence public sentiment and affect government policies. Russia, China, Iran, and other foreign malicious actors all will seek to interfere in the voting process or influence voter perceptions. Adversaries may try to accomplish their goals through a variety of means, including social media campaigns, directing disinformation operations or conducting disruptive or destructive cyber-attacks on state and local infrastructure."

The unusual statement was both a warning and a promise.

Hacking the Vote

U.S. Intelligence agencies concluded China interfered in both the 2016 and 2018 elections, but when confronted with accusations about the 2020 elections, the Chinese Foreign Minister stated China "is not interested" in interfering in the 2020 process. Yet in June 2020, Google announced China was observed targeting former Vice President Joe Biden's campaign and that Iran was targeting President Trump's staffers. It appears there is plenty of interest by foreign actors, state sponsored and otherwise, in affecting the outcome of U.S. elections.

Interference in the election process does not have to originate outside the country. There are plenty of opportunities for agenda-driven groups, political parties, and states to attempt to manipulate the U.S. election process internally. Voter suppression has been an issue in the U.S. since the end of the Civil War. In the 1870s, some southern states instituted poll taxes. Others opted for literacy tests as a way to limit Black voters from reaching the polls. These Jim Crow voting laws weren't eradicated until the Voting Rights Act of 1965.

Getting Out the Vote

Today, according to the American Civil Liberties Union (ACLU), voter suppression ranges from laws requiring voters to present a government ID in order to vote, purges of voter rolls, redistricting efforts to manipulate voting majorities, and systemic disenfranchisements. Women were not enfranchised with the vote until the passage of the 19th Amendment one hundred years ago, in 1920.

Voter fraud. Does it happen? Rarely. Voter impersonation is equally rare. What we hear about are typically ballot errors made by individuals or mistakes by administrators. What we are witnessing now is not the prediction of voter fraud and forgery due to manufactured difficulties with voter registration. It is fear mongering on a national scale, designed to undermine public trust in the process.

In June 2020, President Trump predicted the November presidential election would be "the most corrupt election in the history of our country," claiming there would be widespread voter fraud if people were permitted to vote by mail.

In fact, the move to postal voting has been endorsed by the Center for Disease Control and

Prevention (CDC) as a solid option in the face of the continuing COVID-19 pandemic. States that have long permitted mail-in ballots have reported no issues with widespread fraud. Service members deployed overseas have long voted via absentee ballot. Eleven of the sixteen states that restrict absentee voting, have relaxed the rules to permit absentee ballots for primaries, some for the national elections in November. Yet the fight continues, on a highly partisan basis.

Meantime, the baseless stoking of voter fears concerning mail-in ballots rages on. This is effectively setting the stage for later claims of fraud and establishing a foundation from which to challenge election outcomes. We already know we will hear a litany of "I told you so" false narratives.

Fears about in-person voting have likewise skyrocketed as the pandemic has affected all areas of public life in 2020. People have stood in lines for hours to vote in primaries during the first half of 2020. Some states had to postpone elections — several times — until they could figure out how to keep voters safe during their trip to the polls, others looked for substitutions.

According to Axios, Pennsylvania, Wisconsin, North Carolina and Georgia are at the center of the voting rights issue. "If elections are close in these states and some others, they could set off waves of protests and lawsuits over turnout, ballot access and alleged fraud — and that could undercut the perceived legitimacy of the results."

Voter turnout in the 2018 midterm elections increased sharply. Given the intensity of social justice issues at the forefront of society since that time, increased voter turnout, whether by mail-in ballot or in person, seems certain.

As the 2020 presidential election approaches, we are clearly on a collision course between voter expectations and the process, with a myriad of roadblocks in between. Clearly as the above Axios survey reveals, public expectations are overwhelmingly for an immediate result, despite experts' concern that it may take weeks for the results to be finalized.

The election doesn't have to end with the potential for derailment of trust in the process. There is much that can, and should, be done at all levels. Every single democratic institution has a role to play in securing the election process and it would go a long way towards restoring trust for each of them to pick up the challenge and act.

Protecting the Vote

Institutional response to protecting the election process includes:

- ✓ Federal support to encourage voter registration (it even could be undertaken as part of the census effort) — including providing resources so that there are sufficient polling places and staffing to handle an influx of in-person voters. The result will be shorter wait times and a better voter experience.

- ✓ Transparency in reporting on how absentee or mail-in ballots work.

✓ Polling station staffers must be prepared to ensure the polling places are not overcrowded, social distancing measures are instituted and clearly marked and that all surfaces are continuously cleaned and sanitized. Voters must wear a mask.

✓ The media at all levels can ensure transparency in the vote counting process. Provide advance information on how the process works so that when some city or precinct results take more than one day to tally and confirm, the public sentiment is not rattled by the effects.

✓ Schools, employers, and NGOs can educate the public on how election interference works. The news media should be reporting on how false information is being found and eliminated.

✓ Social media companies should continue to actively hunt down trolls, bots and other sites that spread false narratives and promote conspiracy theories.

✓ Mainstream media should continue to report on how and when threats to the democratic process are discovered and eliminated. Citizens have the right to know what fraud looks like, how to recognize a bot, and how to stop the spread of misinformation.

Will this guarantee that there will not be any issues in the election process? Certainly not. There will always be issues, such as the hanging chad disaster (incompletely punched voting cards) in the 2000 presidential election. But perhaps our biggest challenge today comes from whispers—just a mention of the possibility of election interference can result in what psychology researcher Liv Pelligrini terms "intellectual interference." Those who express fears about potential interference have successfully laid the groundwork for doubt and mistrust to grow unchecked. For once trust has been subjected to intellectual interference, one cannot depend on sanctity of the election process.

All that rational and reasoned voters can do is enact recommended measures to promote and protect the vote. This will go a long way to improve straightforward and truthful communications. As such, they can also be effective at helping to restore confidence in the most basic of democratic processes, the election.

CHAPTER 10

The Road Ahead

Great leaders have a vision of the future that does not yet exist, and an ability to communicate that vision. When we put words to the world we imagine, we can inspire others to join us in creating a brighter future.

—Simon Sinek

The future is replete with opportunities for managing the pace of change, response to change and how information is protected and revealed. Governments are getting tougher on tech companies and are now exerting more influence to prevent the growth of monopolies in information sources. We can expect to see an increasing focus on anti-trust legislation for media and tech corporations and a greater emphasis on the uses of AI to detect disinformation and manipulated content, enforcing protections of both information and privacy of individuals.

Professional standards in journalism are being enforced and medical and legal codes of ethics continue to be refined and internally monitored.

Media companies are beginning to confront the necessity for ensuring free speech meets standards of decency and does not incite illegal behavior.

Education and Insights

There is a critical need for increased education in assessing information for truthfulness and a greater focus on individual, as well as corporate, responsibility in the information domain. With little civics being taught in public schools, by 2016 only twenty-three percent of eighth graders performed at or above the proficiency level on a national-level civics exam.

Of necessity, education is going beyond schools. Tech companies should help users to determine the truthfulness of available information, avoid hackers, and be able to recognize sites and scams with ill intent, all without violating individual rights either to offer opinion or purpose. Numerous police department now provide guides for parents on what social media sites may be inappropriate for their children.

15 Apps Parents Should Know About

Courtesy of the
Madill Police Department

MEETME MEETME is a dating social media app that allows users to connect with people based on geographic proximity. As the app's users are encouraged to meet each other in person.

GRINDR GRINDR is a dating app geared towards gay, bi and transgender people. The app gives users options to chat, share photos and meet up based on a smart phones GPS.

SKOUT SKOUT is a location-based dating app and website. While users under 17 years old are unable to share private photos, kids can easily create an account with an older age.

WHATSAPP WHATSAPP is a popular messaging app that allows users to send texts, photos, voicemails, and make calls and video chats

TIKTOK TIKTOK is a new mobile device app popular with kids used for creating and sharing short videos. With very limited privacy controls, users are vulnerable to cyber bullying and explicit content.

BADBOO BADBOO is a dating and social networking app where users can chat, share photos and videos and connect based on location. While the app is intended for adults only, teens are known to create profiles.

BUMBLE BUMBLE is similar to the popular dating app "Tinder" however, it requires women to make the first contact. Kids have been known to use BUMBLE to create fake accounts and falsify their age.

SNAPCHAT SNAPCHAT is one of the most popular apps in recent years. While the app promises users can take a photo/video and it will disappear, new features including "stories" allows users to view content for up to 24 hours. Snapchat also allows users to see your location.

KIK KIK allows anyone to contact and direct message to your child. Kids can bypass traditional text messaging features. KIK gives users unlimited access to anyone, anywhere, anytime.

LIVEME LIVEME is a live-streaming video app that uses geolocation to share videos so users can find out a broadcaster's exact location. Users can earn "coins" as a way to "pay" minors for photos.

HOLLA HOLLA is a self-proclaimed "addicting" video chat app that allows users to meet people all over the world in just seconds. Reviewers say they have been confronted with racial slurs, explicit content, and more.

WHISPER WHISPER is an anonymous social network that promotes sharing secrets with strangers. It also reveals a user's location so people can meet up.

ASK.FM ASK.FM is known for cyber bullying. The app encourages users to allow anonymous people to ask them questions.

CALCULATOR% CALCULATOR% is only one of SEVERAL secret apps used to hide photos, videos, files, and browser history.

HOT OR NOT HOT OR NOT encourages users to rate your profile, check out people in their area, and chat with strangers. The goal of this app is to hook up.

Source: Medill, Oklahoma Police Department, 2020

In the same vein, the FBI has developed educational programs for middle school students that teach children how to be safe on the internet. The FBI Safe Online Surfing (FBI-SOS) program is a nationwide initiative designed to educate children in grades three through eight about the dangers they face on the internet and to help prevent crimes against children. It promotes cyber citizenship among students by engaging them in a fun, age-appropriate, competitive online program where they learn how to safely and responsibly use the internet. The program emphasizes the importance of cyber safety topics such as password security, smart surfing habits, and the safeguarding of personal information. In the 2019

– 2020 school year more than 1.3 million students took the SOS exam. Since 2012, the FBI's educational site has been visited 12 million times.

These are just initial efforts to stem the tide of declining civics education in the U.S. More needs to be taught, particularly in terms of how to make sense of news and public information. As Joshua Yaffa stated in a recent New Yorker article, "If you don't know how government actually works, you're more likely to believe in conspiratorial versions of its doings."

Universities are quickly pursing their own paths to educating students about democratic processes, information streams and their potential for manipulation. The University of Washington's 2018 online elective "Calling Bullshit," received an overwhelming response from students and educators. The viral clamoring for more knowledge on how to counter misinformation led to the establishment of the Center for an Informed Public. Launched in December 2019, the Center's mission statement is to "resist strategic misinformation, promote an informed society, and strengthen democratic discourse." The Center supports interdisciplinary research, courses, and publications on a broad spectrum of communications issues, ranging from misinformation regarding the COVID-19 pandemic, to the role of bots and trolls in a crisis. Center professor, Dr. Carl Bergstrom commented on the Center's COVID communications research, saying, "This is a crisis unfolding in slow motion, in a statistical way where we can only see pieces of it," he says. "I recommend people pick one maybe two times a day to read what's going on from reputable sources like The New York Times or STAT or WIRED—and if you must go on Twitter, block the hashtags."

Maintaining Engagement

Public engagement with media, has remained strong, growing twenty-two percent in 2019, according to Edelman. Social activism is likewise increasing. More individuals are involved in grassroots movements from #MeToo, to #NeverAgain while Greta Thunberg, the Swedish teenager making a point of how important environmentalism is to the future of the planet has her own book, *No One is too Small to Make a Difference*. Corporate involvement is rising: besides the addressing the issue of climate change at Amazon, Google and Target, other companies have addressed their sexual assault policies and inherent racism in hiring. As regards the issue of immigration, Apple CEO Tim Cook urged the U.S Supreme Court to save the Deferred Action for Childhood Arrivals (DACA) program, protecting hundreds of thousands of young immigrants from deportation. In June 2020 when the Court ruled against the Trump administration's efforts to end the program, Cook said, "The four hundred seventy-eight Dreamers at Apple are members of our collective family. With creativity and passion, they've made us a stronger, more innovative American company. We're glad for today's decision and will keep fighting until DACA's protections are permanent."

Black Lives Matter surged to front of mind for all Americans in the summer of 2020, the season of our discontent. The seemingly nonchalant murder of George Floyd on a city street in Minneapolis broke open a seeping wound of national shame and sparked a major outcry against the police, white supremacy and the full, ugly history of institutional racism across the country. Confederate statues toppled from their bases, the confederate flag was finally folded and America began to confront her troubling past and the issues spawned by institutional racism. At last it seemed that this was the moment—real change was possible. But the spread of the coronavirus also impacted public awareness of racial inequities. According to an Edelman special report published in May, 2020, the spread of COVID-19 highlights the depths of racial disparity across the U.S.

A Gallup poll conducted in August 2020 revealed similar results—that American's view race relations at their lowest point since 2001. The primary rationale given by respondents was the numbers of Black Americans being killed by White police officers.

Focused on Change

As revealed by this latest poll, other changes in public confidence in institutions have taken center stage. This remarkable shift in confidence in institutions is highlighted by the public's growing trust in government to provide information about the pandemic. According to Edelman's 2020 Spring Update, trust in government sources surged eleven percent, with the public relying on government for protection at a level of trust not seen since World War II. Respondents want the government to continue to provide economic relief (eighty-six percent), to get the country back to normal (seventy-nine percent), to contain the spread of the virus (seventy-three percent), and to keep the public informed (eighty-six percent).

There are also great expectations for business to partner with government in improving the economy and building new job opportunities. This represents an inflection point for both business and NGOs. Never has the need been greater for collaborative and cooperative approaches to rebuilding not only the economy, but also the underpinnings of society—

education, access to health care, and opportunities for fair housing, good jobs, and the potential for living a full life. While the May study revealed the pandemic has turned many trust variables around, the need for fair and accurate media coverage continues. The pressures of pandemic response represent a monumental opportunity for traditional news media to regain public trust and confidence. Trust in traditional media grew seven points from January to May 2020 but there is still work to be done.

The media has a major role to play in reporting on the necessity for change in society, as well as promoting education and civic engagement. In order to fulfill its self-described mission to educate, inform and entertain, the broad institution of media must do more to guard information quality, call out false narratives, discipline itself, and protect the privacy of consumers. Media organizations must transform to serve the public, educate and inform, not merely entertain and titillate. Journalists and editors must refrain from glorifying celebrity and adhere to a set of professional standards heretofore neither broadly defined nor enforced. Social media is also displaying increased signs of corporate responsibility. As of November, 2019 Facebook more than tripled its employees working on security and safety. Over thirty-five thousand digital media experts are now focused on AI, content review, machine learning and more.

As the public has become increasingly engaged with news, people are unwilling to merely accept information as it is delivered. Many now question sources, demand transparency, and explore references and sources. One would expect to this to bode well for journalism—greater public involvement should mean the more thorough the fact checking and increased awareness of scams, falsehoods, and hacks. There are more sites that support determining truth in news, both in terms of ferreting out manipulated media—from photographs to videos—and in determining false sources, lies, and bias.

As fallout from the pandemic continues to push democratic institutions, a number of efforts will become more prevalent and visible:

✓ Increasing focus on regulation of tech firms, anti-trust cases and a major focus on corporate responsibility for safeguarding privacy and truth.

✓ New respect for the planet, environment issues, awareness of the effects of global warming and the creative abilities of science.

✓ Increased government and business efforts to protect privacy, stop hackers, find and force out bots and trolls, delete false accounts, and call out those who incite violence.

✓ Growing transparency in intelligence and awareness that shared information can effectively prevent false narratives and influence campaigns.

✓ New applications for AI and the growth of facial recognition software applications.

✓ Continuing discussion in the diplomatic arena of how war in cyberspace can result in war in the physical domains. The second and third order effects concerning cyber rules of engagement and legal rulings regarding justification form declarations of war will dominate military discussions.

These actions represent the signs of a coming transformation, a culture shift from the Information Age's "apocalyptic Wild West" to tomorrow's Knowledge Age. In this next generation of information, trust will be paramount. The responsibility for creating this massive cultural change lies with everyone: governments, business, NGOs, and the media—in short with all institutions that have a stake in preserving trust and truth, and with all citizens. Thankfully, a harsh and dystopian future isn't a guaranteed outcome. We need to consciously develop media literacy standards in education at all levels, mandate responsible media, and build public trust and confidence in the knowledge that the false will be found out and eliminated.

These transformative actions are significant in and of themselves, but it is critical to go further. According to futurist Anne Lise Kjaer, "Things no longer change over a generation or a decade, but from year to year, even month to month, creating new arenas for disruptive ideas and innovation to emerge...Inevitably, this leads us to reconsider the future and our place within it." We must act in order to make sense of the rapid pace of change, today and tomorrow, developing not just a media literacy but futures literacy.

Many organizations are going even further, determined to meet future challenges head on. These companies are scheduling foresight sessions, hosted by futurist companies like the Institute for the Future. The Institute and like organizations train business and government organizations to identify disruptors, recognize trends, develop opportunities for change management, and scout out indications of emerging possibilities. Today's volatile operating environment has leaders realizing the key to prevailing in an uncertain future lies in the ability to not only identify potential change but to adapt in advance. But at the most basic level, responsibility for this change begins with the individual.

BUILDING TRUST
FOR THE FUTURE

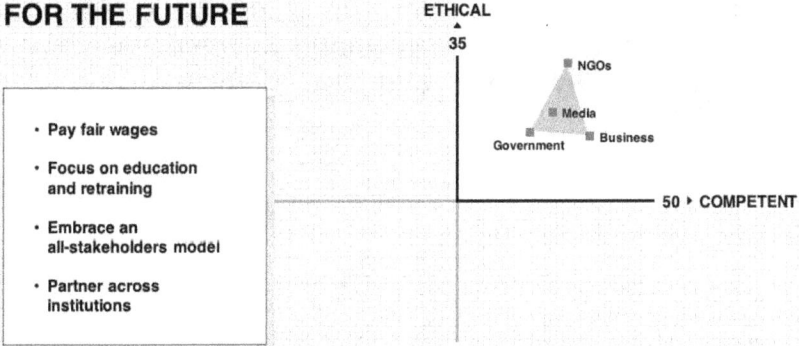

ETHICAL

35

■ NGOs

■ Media

Government ■   ■ Business

50 ▸ COMPETENT

- Pay fair wages

- Focus on education and retraining

- Embrace an all-stakeholders model

- Partner across institutions

2020 Edelman Trust Barometer. For details regarding how this model of trusted institutions, please see the Technical Appendix.

Citizens must take measures to secure and protect their privacy online and in the public space. Education must become more robust and available at all levels, not just in schools. Leaders must drive change within their organizations and cultures. While many experts predict digital life in the future will improve, they also caution that people must be engaged in the process to effect this transformation. Foresight increases agility and the ability to adapt and ensures that the fears many hold today are fully addressed. Basic rights and economic fairness must be taken into account and watched carefully.

With effort, tomorrow's social media platforms will be increasingly socially responsible and committed to protecting the privacy and rights of its users. Future citizens will undoubtedly be better informed because they can have trust and confidence in the knowledge they possess, know their privacy is secure, and can trust their institutions. The Knowledge Age is there to be created, shaped, and developed by today's tech firms, policy experts, and citizen activists. Everyone must embrace these innovations and reforms, insist on basic rights, fairness, and demand factual information that we can all trust, rely upon and use with the utmost confidence.

AFTERWORD

Jacqueline F. Strayer

It is the mark of an educated mind to be able to entertain a thought without accepting it.

—Aristotle

Trust is foundational to healthy relationships. We come into the world hardwired as social creatures and have an innate tendency for social connections and to affiliate with others. This need to trust extends not only in our personal relationships, which we develop in our early years, but later in our professional lives, in the organizations we work for, the leaders who govern us and the societal institutions of which we are members. Our trust extends to the information we consume, the products we buy and the way we think about our futures.

We cannot live our lives in comfort if we cannot trust others. And yet we know <u>trust is dynamic</u>. We trust one day, something happens and all of a sudden, we are disappointed, our trust has been violated and we are mistrustful. And now we find ourselves in a very different world than just a few months ago, living in a global pandemic and all of the sudden our concept of trust has taken on different dimensions and new meaning.

Research provides us with a window of what we are experiencing. The state of our personal trust with each other is under attack. The Pew Center reported that, even before the pandemic, many Americans are anxious about the level of confidence they have in each other. In fact over seventy percent believe that interpersonal confidence has declined in the last two decades.

The unprecedented crisis we are living in challenges societal norms of trust. It influences out ability to trust each other to do the right thing and how we think about our own safety. If we do not believe that others will play by the restrictive rules that have been established during this pandemic, then we are less likely to play by them ourselves. This is derived from the "problem of collective action, which arises when actions that would lead to the common good seemingly conflict with those that advantage the individual" (Rothstein 2020).

It will be years until we will know the effect the COVID-19 pandemic has had on our global community and while trust in our institutions has been under siege, the pandemic has globally shifted some of our trust, where we place it, and in surprising ways.

The Global 2020 Edelman Trust Barometer update in May 2020 pointed out that the pandemic caused significant change in trust and reputation for major institutions. This is distinct from their annual survey published in January 2020 which showed very different results from their updated survey. In this Spring Update, it indicated that amid the COVID-19 pandemic, government trust surged eleven points to an all-time high of sixty-five percent, making government the most trusted institution for the first time in their twenty years of study.

This research underscores that events can trigger and change our perception of what we trust, when and the extent of it. For example, trust in the media has soared during this time, as we sought out information, both through paid (up eight percent in this recent survey) and earned (up seven percent) media. However societal inequities were highlighted by their research. Their Spring Update shows that "sixty-seven percent of respondents believe that those with less education, less money and fewer resources are bearing a disproportionate burden of the suffering, risk of illness and need to sacrifice in the pandemic, and more than half are very worried about long-term, COVID-related job loss." These inequities will undoubtedly influence how we view longer-term the response to these hard-hit parts of our global community and our trust of these institutions in our evaluation of their response.

Two other recent studies illuminate other issues the pandemic has created. One study found that due to the current recession and likely consequences of it, fear has been created that may have long lasting implications for businesses and organizations to resume previous spending patterns, thereby creating ongoing economic erosion. And, it shows that people who endure the pandemic in young adulthood may be more distrustful of government institutions for the rest of their lives, an outcome that could make it more difficult for governments to effectively respond to future pandemics, when or if, they occur.

So, while trust seems to be building globally for certain institutions (even business and NGOs are up four percent), we are still in the midst of the pandemic at this writing and cannot make definitive statements or predictions. It is unclear when we will come out the other side of it what the result will be, the consequences and how trust will be viewed of government, the media and the business community.

Jacqueline Strayer is a former elected officer for three Fortune 500 companies, overseeing marketing and public relations. Today she serves as faculty member in graduate and executive programs at New York University and Columbia University. She writes about contemporary issues and their relationship to leadership, marketing and the public relations practice.

AUTHOR'S NOTE

The majority of the chapters here originally appeared as separate articles in a series on the Information Apocalypse, published by the U.S. Army War College's WarRoom. They have been updated and reformatted for this publication.

The appearance of U.S. DoD visual information does not imply or constitute DoD endorsement.

Graphics attributed to the Pew Research Center are included according to the terms of use as established by the Pew Research Center, Washington, D.C.

Graphics attributed to the Edelman Trust Barometer Report (2018, 2019, 2020) are provided courtesy of Edelman.com and have not been changed or modified from the Edelman-produced originals.

FURTHER READING

Books

Clapper, James R. with Brown, Trey. *Facts and Fears: Hard Truths from a Life in Intelligence.* New York: Viking/Penguin Random House LLC, 2018.

Farman, Jason. *Delayed Response: The Art of Waiting from the Ancient to the Instant World.* New Haven, CT: Yale University Press, 2018.

Friedman, George. *The Next 100 Years: A Forecast for the 21st Century.* New York: Anchor Books/Random House Inc., 2009.

Galloway, Scott. The Four: *The Hidden DNA of Amazon, Apple, Facebook, and Google.* New York: Portfolio/Penguin, 2017.

Iger, Robert. *The Ride of a Lifetime: Lessons Learned from 15 Years as CEO of the Walt Disney Company.* New York: Random House, 2019.

Kakutani, Michiko. *The Death of Truth: Notes on Falsehood in the Age of Trump.* New York: Tim Duggan Books, 2018.

Levin, Yuval. *A Time to Build: From Family and Community to Congress and the Campus, How Recommitting to Our Institutions Can Revive the American Dream*. New York: Hachette Books, 2020.

McGonigal, Jane. *Reality is Broken: Why Games Make Us Better and How They Can Change the World*. New York: Penguin, 2011.

McIntyre, Lee. *Post Truth*. Cambridge, MA: The MIT Press, 2018.

Patrikarakos, David. *War in 140 Characters: How Social Media is Reshaping Conflict in the Twenty-First Century*. New York: Basic Books, 2017.

Rubin, James and Carmichael, Barie. *Reset: Business and Society in the new Social Landscape*. New York: Columbia University Press, 2017.

Sinek, Simon. *Start With Why: How Great Leaders Inspire Everyone to Take Action*. New York: Portfolio/Penguin, 2009.

Singer, P.W. *Wired for War: The Robotics Revolution and Conflict in the 21st Century*. New York: Penguin Books, 2009.

Singer, P.W. and Brooking, Emerson T. *Like War: The Weaponization of Social Media*. New York: Houghton Mifflin Harcourt, 2018.

Stengel, Richard. *Information Wars: How We Lost the Global Battle Against Disinformation & What We Can Do About It.* New York: Atlantic Monthly Press, 2019.

Special Edition Periodicals

"Truth, Lies & Uncertainty: Seaching for Reality in Unreal Times," *Scientific American*, Vol. 321, #3, September 2019.

"World War Web: The Fight for the Internet's Future," *Foreign Affairs*, Vol. 97, #5, September/October 2018.

Websites

Center for an Informed Public, the University of Washington. https://www.cip.uw.edu/

The Edelman Trust Barometer
https://www.edelman.com/research/

Hidden Tribes, A Study of America's Polarized Landscape.
https://hiddentribes.us/pdf/hidden_tribes_report.pdf

Pew Research Center.
https://www.pewresearch.org/

REFERENCE LIST

2018 Edelman TRUST BAROMETER (2018) *Edelman.com*. Available at: https://www.edelman.com/research/2018-edelman-trust-barometer (Accessed: November 16, 2020).

American Experience (no date) *World War II Propaganda, Pbs.org*. Available at: https://www.pbs.org/wgbh/americanexperience/features/goebbels-propaganda/ (Accessed: November 16, 2020).

American Historical Association (1944) *The Story of Propaganda, Historians.org*. Available at: https://www.historians.org/about-aha-and-membership/aha-history-and-archives/gi-roundtable-series/pamphlets/em-2-what-is-propaganda-(1944)/the-story-of-propaganda (Accessed: November 16, 2020).

Baer, B. (2019) *Video: Astros relay signs to hitters by banging on a trash can in 2017, Nbcsports.com*. Available at: https://mlb.nbcsports.com/2019/11/12/video-astros-relay-signs-to-hitters-by-banging-on-a-trash-can-in-2017/ (Accessed: November 16, 2020).

Boyd, D. (2011) *Compassion fatigue - the American institute of stress, Stress.org*. Available at: https://www.stress.org/military/for-practitionersleaders/compassion-fatigue (Accessed: November 16, 2020).

Brittain, A. (2018) "Charlie Rose's misconduct was widespread at CBS and three managers were warned, investigation finds," *Washington post (Washington, D.C.: 1974)*, 2 May. Available at: https://www.washingtonpost.com/charlie-roses-misconduct-was-widespread-at-cbs-and-three-managers-were-warned-investigation-finds/2018/05/02/80613d24-3228-11e8-94fa-32d48460b955_story.html (Accessed: November 16, 2020).

Calling Bullshit (no date) *Callingbullshit.org*. Available at: https://www.callingbullshit.org/ (Accessed: November 16, 2020).

Carvajal, E. (2020) *Arnold Schwarzenegger confirmed that Maria shiver knew about his affair for 'many years,' news.amomama.com*. Available at: https://news.amomama.com/190268-arnold-schwarzenegger-confirmed-that-mar.html (Accessed: November 16, 2020).

Cassidy, J. (2019) "Alex Acosta had to go, but the Jeffrey Epstein scandal is really about money and privilege," *New Yorker (New York, N.Y.: 1925)*, 12 July. Available at: https://www.newyorker.com/news/our-columnists/alex-acosta-had-to-resign-but-the-

epstein-scandal-goes-well-beyond-his-role
(Accessed: November 16, 2020).

Corn, D. (2020) "Newly released transcripts show
Michael Flynn betrayed the United States," *Mother
Jones*, 29 May. Available at:
https://www.motherjones.com/politics/2020/05/newl
y-released-transcripts-show-michael-flynn-betrayed-
the-united-states/ (Accessed: November 16, 2020).

Ellis, B. (2016) "Guides: Internet news, fact-
checking, & critical thinking: Non-partisan fact
checking sites." Available at:
https://middlebury.libguides.com/Internet/fact-
checking (Accessed: November 16, 2020).

Flood, B. (2020). *Facebook's Fall: From the Friendliest
Face of Tech to Perceived Enemy of Democracy. Fox
News*. Available at:
https://www.foxnews.com/tech/facebook-went-
from-happy-face-of-tech-to-perceived-enemy-of-
democracy (Accessed: November 16, 2020).

Fox, E. J. (2017) *Brian Williams opens up about his
unexpected re-invention: "second acts are possible, with a
little spiffing up," Vanity Fair*. Available at:
https://www.vanityfair.com/news/2017/10/brian-
williams-11th-hour (Accessed: November 16, 2020).

Gallup, Inc. (2020) *Confidence in Institutions,
Gallup.com*. Gallup. Available at:
https://news.gallup.com/poll/1597/confidence-
institutions.aspx (Accessed: November 16, 2020).

Gecewicz, C. (2019) *Key takeaways about how Americans view the sexual abuse scandal in the Catholic Church, Pewresearch.org*. Available at: https://www.pewresearch.org/fact-tank/2019/06/11/key-takeaways-about-how-americans-view-the-sexual-abuse-scandal-in-the-catholic-church/ (Accessed: November 16, 2020).

General Data Protection Regulation (GDPR) compliance guidelines (2018) *Gdpr.eu*. Available at: https://gdpr.eu/ (Accessed: November 16, 2020).

Goldman, A. *et al.* (2020) "Lawmakers are warned that Russia is meddling to re-elect trump," *The New York times*, 20 February. Available at: https://www.nytimes.com/2020/02/20/us/politics/russian-interference-trump-democrats.html (Accessed: November 16, 2020).

Green, J. (2018) "Review: A Three-Way smackdown over 'the lifespan of a fact,'" *The New York times*, 19 October. Available at: https://www.nytimes.com/2018/10/18/theater/lifespan-of-a-fact-review-daniel-radcliffe-bobby-cannavale.html (Accessed: November 16, 2020).

Hubbard, L. (2018) *Here's what Matt Lauer is doing now, Town & Country*. Available at: https://www.townandcountrymag.com/society/a15845582/what-matt-lauer-is-doing-now/ (Accessed: November 16, 2020).

Josephson, M. S. (2005) *Preserving the public trust: The five principles of public service Ethics.* Unlimited Publishing.

Kakutani, M. (2018) "James Comey Has a Story to Tell. It's Very Persuasive," *The New York Times*, 12 April. Available at: https://www.nytimes.com/2018/04/12/books/review/james-comey-a-higher-loyalty.html (Accessed: November 16, 2020).

Ma, A. and Gilbert, B. (2019) "Facebook understood how dangerous the Trump-linked data firm Cambridge Analytica could be much earlier than it previously said. Here's everything that's happened up until now," *Business Insider*, 23 August. Available at: https://www.businessinsider.com/cambridge-analytica-a-guide-to-the-trump-linked-data-firm-that-harvested-50-million-facebook-profiles-2018-3 (Accessed: November 16, 2020).

Marcus, J. (2020) "Quarantine Fatigue Is Real," *Atlantic monthly (Boston, Mass.: 1993)*, 11 May. Available at: https://www.theatlantic.com/ideas/archive/2020/05/quarantine-fatigue-real-and-shaming-people-wont-help/611482/ (Accessed: November 16, 2020).

Matsa, K. E. (2018) *News use across social media platforms 2018, Journalism.org.* Available at: https://www.journalism.org/2018/09/10/news-use-across-social-media-platforms-2018/ (Accessed: November 16, 2020).

Media Bias Chart - ad Fontes media (2019) *Adfontesmedia.com*. Available at: https://www.adfontesmedia.com/interactive-media-bias-chart/ (Accessed: November 16, 2020).

Mitchell, A. (2018) *Americans want online information freedoms over government restriction of fake news, Journalism.org*. Available at: https://www.journalism.org/2018/04/19/americans-favor-protecting-information-freedoms-over-government-steps-to-restrict-false-news-online/ (Accessed: November 16, 2020).

Mohsin, M. (2020) "Top 10 Facebook statistics you need to know in 2020," *Oberlo.com*. Oberlo, 10 May. Available at: https://www.oberlo.com/blog/facebook-statistics (Accessed: November 16, 2020).

Mueller, R. (2019) *Report on The Investigation Into Russian Interference in the 2016 Presidential Election, Justice.gov*. Available at: https://www.justice.gov/storage/report.pdf (Accessed: November 16, 2020).

New ICD-10-CM code for the 2019 Novel Coronavirus (COVID-19 (2020) *Cdc.gov*. Available at: https://www.cdc.gov/nchs/data/icd/Announcement-New-ICD-code-for-coronavirus-3-18-2020.pdf (Accessed: November 16, 2020).

The New York Times (no date) "Roman Catholic Church Sex Abuse Cases." Available at: https://www.nytimes.com/topic/organization/roman

-catholic-church-sex-abuse-cases (Accessed: November 16, 2020).

Noor, I. (2020) *Confirmation Bias, Simplypsychology.org*. Available at: https://www.simplypsychology.org/confirmation-bias.html (Accessed: November 16, 2020).

Orwell, G. (1944) *Looking Back on the Spanish War, Orwellfoundation.com*. Available at: https://www.orwellfoundation.com/the-orwell-foundation/orwell/essays-and-other-works/looking-back-on-the-spanish-war/ (Accessed: November 16, 2020).

Ovadya, A. (2018) *We are careening toward a future, Twitter.com*. Available at: https://twitter.com/metaviv/status/963188182861864960 (Accessed: November 16, 2020).

Panja, T. (2019) "Russia banned from Olympics and global sports for 4 years over doping," *The New York Times*, 9 December. Available at: https://www.nytimes.com/2019/12/09/sports/russia-doping-ban.html (Accessed: November 16, 2020).

Patterson, D. (2020) *Facebook data privacy scandal: A cheat sheet, TechRepublic*. Available at: https://www.techrepublic.com/article/facebook-data-privacy-scandal-a-cheat-sheet/ (Accessed: November 16, 2020).

Paulson, M. (2020) "Broadway, symbol of New York resilience, shuts down amid virus threat," *The*

New York times, 12 March. Available at: https://www.nytimes.com/2020/03/12/theater/corona virus-broadway-shutdown.html (Accessed: November 16, 2020).

Public Affairs Pulse Survey - Public Affairs Council (2020) *Pac.org*. Available at: https://pac.org/public-affairs-pulse-survey/ (Accessed: November 16, 2020).

Quell, M. (2020) *More Countries Pass 'Fake News' Laws in Pandemic Era, Courthousenews.com*. Available at: https://www.courthousenews.com/more-countries-pass-fake-news-laws-in-pandemic-era/ (Accessed: November 16, 2020).

Ransom, J. (2020) "Harvey Weinstein's stunning downfall: 23 years in prison," *The New York Times*, 11 March. Available at: https://www.nytimes.com/2020/03/11/nyregion/harv ey-weinstein-sentencing.html (Accessed: November 16, 2020).

Reddy, N. (2019) *Harassment, accountability, and the erosion of judicial legitimacy, Americanprogress.org*. Available at: https://www.americanprogress.org/issues/courts/ne ws/2019/08/05/473160/harassment-accountability-erosion-judicial-legitimacy/ (Accessed: November 16, 2020).

Rogin, J. (2013) *Exclusive: McChrystal was shocked by controversy over Rolling Stone article, Foreign Policy*. Available at:

https://foreignpolicy.com/2013/01/04/exclusive-mcchrystal-was-shocked-by-controversy-over-rolling-stone-article/ (Accessed: November 16, 2020).

Rothstein, B. (2020). Truth is the Key to Fighting the Pandemic. *Scientific American.* Available at: https://blogs.scientificamerican.com/observations/trust-is-the-key-to-fighting-the-pandemic/ (Accessed: November 16, 2020).

Rouse, M. (2020) *What is Artificial Intelligence (AI)?, Techtarget.com.* TechTarget. Available at: https://searchenterpriseai.techtarget.com/definition/AI-Artificial-Intelligence (Accessed: November 16, 2020).

Schake, K. (2018). *Social Media as War? War on the Rocks.* Available at: https://warontherocks.com/2018/09/social-media-as-war/ (Accessed: November 16, 2020).

Schudson, M. (2019) *The fall, rise, and fall of media trust, Cjr.org.* Available at: https://www.cjr.org/special_report/the-fall-rise-and-fall-of-media-trust.php (Accessed: November 16, 2020).

Seifert, K. (2020) *What really happened during Deflategate? Five years later, the NFL's "scandal" aged poorly, ESPN.* Available at: https://www.espn.com/nfl/story/_/id/28502507/what-really-happened-deflategate-five-years-later-nfl-scandal-aged-poorly (Accessed: November 16, 2020).

Shane, L. (2019) *Survey: Public confidence in the military is high, especially among older generations, Military Times*. Available at: https://www.militarytimes.com/news/pentagon-congress/2019/07/22/survey-public-confidence-in-the-military-is-high-especially-among-older-generations/ (Accessed: November 16, 2020).

Sherman, S. (2017) Facebook is the Enemy Now. *The Huffington Post*. Available at: https://www.huffpost.com/entry/facebook-is-the-enemy-now_b_5a396166e4b0860bf4ab9586 (Accessed November 16, 2020).

Spector, J. (2018) "Behind the scenes: The Capitol chaos when Eliot Spitzer resigned 10 years ago," *Democrat & Chronicle*, 9 March. Available at: https://www.democratandchronicle.com/story/news/politics/albany/2018/03/08/behind-scenes-capitol-chaos-when-eliot-spitzer-resigned-10-years-ago/404266002/ (Accessed: November 16, 2020).

Stucky, G. (2016) "LibGuides: Fake news or real? or how to become media savvy: Confirmation bias." Available at: https://bethelks.libguides.com/c.php?g=591268&p=4194631 (Accessed: November 16, 2020).

Tapper, J. and Correspondent, C. W. (2012) "Events leading to Petraeus resignation," *CNN*, 14 November. Available at: https://www.cnn.com/videos/bestoftv/2012/11/13/ac-petraeus-affair-timeline.cnn (Accessed: November 16, 2020).

Ungarino, R. (2019) "Kylie Jenner's tweet that whacked Snap's stock was one year ago — and shares have never really recovered (SNAP)," *Business Insider*, 21 February. Available at: https://markets.businessinsider.com/news/stocks/snap-stock-price-kylie-jenner-snapchat-tweet-year-ago-2019-2-1027973126 (Accessed: November 16, 2020).

US National Security Advisor LTG H.R. McMaster: "Russian aggression is strengthening our resolve" - Atlantic Council (2018) *Atlanticcouncil.org*. Available at: https://www.atlanticcouncil.org/commentary/transcript/us-national-security-advisor-lt-gen-h-r-mcmaster-russian-aggression-is-strengthening-our-resolve/ (Accessed: November 16, 2020).

Wallace, C. (2016) *Obama did not ban the pledge*, *Factcheck.org*. Available at: https://www.factcheck.org/2016/09/obama-did-not-ban-the-pledge/ (Accessed: November 16, 2020).

Westerheide, F. (2020) "China – the first artificial intelligence superpower," *Forbes Magazine*, 14 January. Available at: https://www.forbes.com/sites/cognitiveworld/2020/01/14/china-artificial-intelligence-superpower/ (Accessed: November 16, 2020).

What are reliable sources for fact-checking and recognizing "fake news"? - Find More Answers (no date) *American.edu*. Available at: https://answers.library.american.edu/faq/282165 (Accessed: November 16, 2020).

Wikipedia contributors (2020a) *Brett Kavanaugh Supreme Court nomination, Wikipedia, The Free Encyclopedia.* Available at: https://en.wikipedia.org/w/index.php?title=Brett_Kavanaugh_Supreme_Court_nomination&oldid=987807847 (Accessed: November 16, 2020).

Wikipedia contributors (2020b) *COVID-19 pandemic, Wikipedia, The Free Encyclopedia.* Available at: https://en.wikipedia.org/w/index.php?title=COVID-19_pandemic&oldid=988998630 (Accessed: November 16, 2020).

Willow Research (2019) *Do Americans have confidence in the courts?, Willowresearch.com.* Available at: https://willowresearch.com/american-confidence-courts/ (Accessed: November 16, 2020).

ABOUT THE AUTHOR

Mari K. Eder, retired U.S. Army Major General, is a renowned speaker and author, and a thought leader on strategic communication and leadership. General Eder has served as Director of Public Affairs at the George C. Marshall European Center for Security Studies and as an adjunct professor and lecturer in communications and public diplomacy at the NATO School and Sweden's International Training Command. She speaks and writes frequently on communication topics in universities and for international audiences and consults on communications issues.

When not writing, speaking, or traveling, she works with rescue groups and fosters dogs for adoption.

www.ingramcontent.com/pod-product-compliance
Lightning Source LLC
Chambersburg PA
CBHW062059270326
41931CB00013B/3138